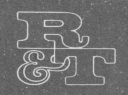

ROAD & TRACK

ON
CORVETTE
1953-1967

Reprinted From
Road & Track Magazine

ISBN 0 907 073 719

Published By
Brooklands Books with permission of Road & Track

Titles in this series

ROAD & TRACK ON FERRARI 1968-1974
Road Tests — New Model Reports — Racing Classic Road Tests
Owner Surveys — Salon — Comparison Tests — European Tests — Track Tests
330 GTS — P4 — 365 GTB/4 — 321P — 340 Mexico Berlinetta — 365 GT 2+2
250 GT Lusso — 330 GTS Targa — 312B — 246 GT — 265 GTC4 — 166 MM
GTO — Dino 308 GT4

ROAD & TRACK ON FERRARI 1975-1981
Road Tests — Styling — European Tests — New Model Reports — Analysis
Track Tests — Salons — Comparison Tests — 308 GT4 — 308 GTB — 308 GTS
308 GTi — 166 Inter — 340 Coupé — F1 312 B3 — 512 BB — 365 GT4
Berlinetta Boxer

ROAD & TRACK ON FIAT SPORTS CARS 1968-1981
Road Tests — New Model Reports — Model Analysis — Owner Impressions
Styling Analysis — Used Car Classics — Owner Surveys — Driving Impressions
Comparison Tests — 850 Spider — 850 Coupé — 124 Sport Coupé — 124 Spider
— 124 Sport Spider — 124 Spider 1600 — Spider 2000 — Spider Turbo
X1/9 — X1/9 1500 5 Speed — X1/9 Fuel Injected

ROAD & TRACK ON MERCEDES SPORTS & GT CARS 1970-1980
Road Tests — Driving Impressions — Salons — Styling Analysis — New Model
Reports — Technical Analysis — Comparison Tests — 450SL — 450SLC 5.0
350SL — 350SLC — 350SL 4.5 — C111 — C111 Mk2 — W154 — Model S
500K — 540K — Cw311 — 300SLR

ROAD & TRACK ON CORVETTE 1953-1967
Road Tests — New Model Reports — Owner Survey — Styling —
Comparison Tests — 36,000 Mile Report — Technical Analysis — Classic Test
SS — Six — V8 — 236/150 — 265/195 — 283/283 — 283/290 — 283/315 —
Sting Ray — 327/300 — 327/360 — 327/375 — 396/425

ROAD & TRACK ON CORVETTE 1968-1982
Road Tests — History — New Model Reports — Comparison Tests — Sting Ray
5-Speed — 4-Door — Prototypes — Cross Fire Injection — LT-1 — 327/350
427/435 — 454/390 — 350/250 — 467/700 — 350/210 — 350/190-142
350/200-149

Titles in preparation will cover:—
Porsche, Lamborghini, Jaguar, BMW etc.

Distributed By

Road & Track
1499 Monrovia,
Newport Beach,
California 92651, U.S.A.

Brooklands Book Distribution Ltd.
Holmerise, Seven Hills Road,
Cobham, Surrey KT11 1ES,
England

Contents

We are frequently asked for copies of out of print Road Tests and other articles that have appeared in Road & Track. To satisfy this need we are producing a series of books that will include, as nearly as possible, all the important information on one make or subject for a given period.

It is our hope that these collections of articles will give an overview that will be of value to historians, restorers and potential buyers, as well as to present owners of these automobiles.

Chevrolet Corvette

The Chevrolet Corvette will become the first volume-produced sports car made in America in over a decade.

Mr. T. H. Keating, Chevrolet general manager has announced that sample cars will be in the hands of principal dealers in 1953 and that about 3,000 units will be produced during 1954. It has also been stated that the Corvette is not intended to be a "racing sports car", but Jaguar said the same thing in 1949. Any sports car addict knows that if you can't race it, *it isn't a sports car!* The Corvette will be used in competition and it has every chance of its share of successes.

Trade gossip says the price will be about $3500 and it is an open secret that the entire contemplated production is "sold."

The powerplant is a modified 1953 assembly which has full pressure lubrication and aluminum pistons. The power output has been stepped up to 160 bhp at 5200 rpm by virtue of higher compression ratio, a special camshaft, side draft Carter carburetors, and a dual exhaust system.

The specification of a Powerglide transmission has met with considerable derision but a torque convertor has potential advantages for road racing which have not been fully explored.

The chassis is specially designed for this car and one notes that Hotchkiss drive is used, a drastic step for Chevrolet. There is independent front suspension and a spe-cial gear ratio of 3.27 is specified. The disc-wheels and ELP tires are not very functional, but if the number of accessories now available for the ubiquitous MG is any indication, this problem can be solved.

A close examination of the fiberglass body reveals excellent workmanship and careful attention to detail. The fabric top is much better in appearance than most imported cars and folds into a flush compartment behind the bucket type seats. The instrument panel is very neat and includes a tachometer placed in the center of the dash.

An estimate of the performance capabilities of the Corvette may be interesting —though premature. Top speed, 120 mph

The Corvette seats are well designed to eliminate "sliding about" when the car is maneuvered.

The Corvette engine. Cylindrical tank, just above the three side draft carburetors, functions as the radiator top tank, necessitated by the extremely low hood line.

Configuration is neat, although some criticism may be directed at the lack of bumpers.

at 4900 rpm; acceleration from zero to 60 mph, 11 seconds; standing ¼ mile in 18 seconds. The *Road and Track* performance index (based on a Cad-Allard = 100) is only 50.00 if our assumption of 195 ft/lbs. of torque is correct. This low figure may account for the employment of a torque multiplying convertor in the transmission. It also makes the possibility of installing a 302 cu in. GMC engine attractive in view of the present general use of F.I.A. regulations where Class C includes displacements from 183 to 305 cu in. —I.B.

Specifications

General: Wheelbase, 102 in.; Tread, front 57 in.; rear 59 in.; tire size, 6.70 x 15; curb weight, 2,900 lbs. approx.

Engine: Chevrolet, 6 cylinder ohv; 3.562 x 3.937; 235.5 cu. in. (3861 cc) 160 bhp at 5200 rpm. Torque (estimated), 195 ft/lbs at 2400 rpm. Three Carter carburetors. Dual exhaust system.

Transmission: Chevrolet Powerglide with floor mounted control. Overall ratios 3.27 and 5.97 plus torque convertor multiplication of 2.1 at stall.

Rear Axle: Hypoid gears with 3.27 to 1 ratio. Open driveshaft.

Suspension: I.F.S. with wishbones and coil springs. Rear, semi-elliptic leaf springs.

Brakes: Bendix duo-servo hydraulic with 11 inch drums.

Dimensions: Overall length 167 in.; width, 70 in., height to top of door, 33 in.

Clean functional lines of the Corvette reflect the fact that this is a genuine sports car, a refreshing contrast to the pseudo sports cars being shown by other divisions of GM.

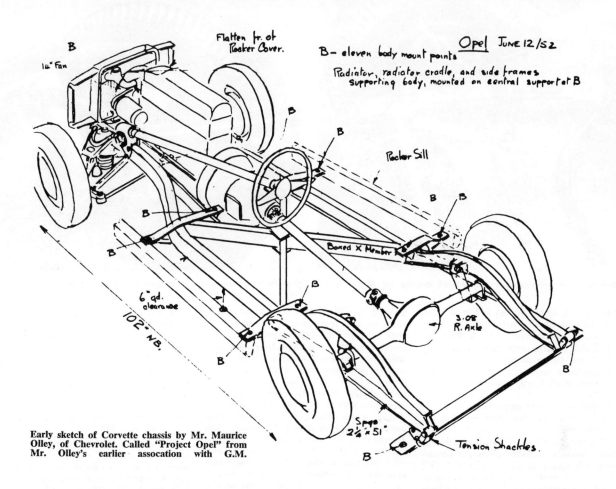

Early sketch of Corvette chassis by Mr. Maurice Olley, of Chevrolet. Called "Project Opel" from Mr. Olley's earlier assocation with G.M.

THE CHEVROLET CORVETTE

by John R. Bond

photographs courtesy Chevrolet Motor Division

As far back as 1948, *Road & Track* took-up the cause of the American Sports Car, with editorials and articles continuing through the years. Now we have such a car in the Chevrolet Corvette· During a recent trip to Chicago and New York it was discovered that very few sports car enthusiasts know much about this new car. There is also a very general feeling that the Corvette is not a genuine sports car. Maybe it isn't, though part of the difficulty can be traced to the lack of a universally accepted definition.

To enable each reader to evaluate the Corvette for himself, the following facts and information are presented. The basis for this report has been taken from an S.A.E. paper, presented by Mr. Maurice Olley, at Detroit, Michigan, on October 5, 1953.

The thinking behind the Chevrolet Corvette is based on the assumption that a sports car must have a cruising speed of over 70 mph, a weight/power ratio of better

(Continued on next page)

Final result, a well engineered sports car chassis.

Development of the new car enjoyed the advantage of G.M.'s vast proving grounds.

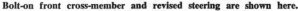
Bolt-on front cross-member and revised steering are shown here.   Corvette rear suspension with special springs and hotchkiss drive.

than 25 to 1, ample brakes, and good handling qualities. On this latter point Mr. Olley lists the following items as being desirable:

1. Quick steering with light handling.
2. A low center of gravity.
3. Minimum overhang with a low moment of inertia relative to wheelbase.
5. Smooth yet firm suspension.
6. A quick steering response, but no oversteer.

How the Chevrolet Motor Division achieved these obviously desirable sports car characteristics is very interesting. On June 2, 1952 Chevrolet engineers were shown a plaster model of a proposed car having a wheelbase of 102 inches. In seven months they were to design, build and test a chassis, using components of known reliability. It must have adequate performance, a comfortable ride and stable handling qualities.

On June 12, one of several sketches was

Manifold side of the 150 bhp Chevrolet engine. Flattened rocker cover allows low hood.

selected (see illustration). It remains fairly close to the final design. In general the chassis components were adapted from Chevrolet parts, but a hotchkiss drive was essential, since the short wheelbase would have required a torque tube so short as to produce excessive change of wheel speed on rough roads. Since the open body would contribute nothing to overall rigidity, a completely special frame was designed using box-section side rails and X-member. The X-member is low enough to allow the drive line to run above it, giving a very strong, solid junction at the "X". The low frame also required outboard rear spring mountings which places them close to the wheels, for stability. Weight of the complete frame is 213 lbs.

The front suspension uses many standard parts but is stiffer in roll by virtue of a larger diameter stabilizer bar. The coil springs are special because of the reduced

load, but their rate appears to be the same as the stock sedan. However the sprung weight is less than stock, giving the effect of "stiffer" springs with a faster bounce frequency. Static deflection is given as 7 inches, equivalent to a ride rate of about 105 lbs/in. and a bounce frequency of 71 oscillations per min. The front cross member appears to be stock and retains the excellent bolt-on, sub-assembly feature used by Chevrolet since 1934.

The steering idler arm (see photo) was redesigned because of the lower engine mounting and is carried on a double-row ball bearing. Steering ratio selected is 16 to 1. On this Mr. Olley says "We are aware of a preference in some quarters for a rack and pinion steering on cars of this type. However this involves a steering ratio of the order of 9 or 10 to 1. We regard this as too fast even for a sports car . . ." The steering wheel is 17.25" in dia., its angle is 13° from vertical and the turning circle is 38 ft.

The rear springs are 2" wide x 51" long, with four leaves. They are inclined (low in front, high at the rear) so as to give approximately 15% roll understeer. Quoting Mr. Olley, "This may appear excessive, but some of the handling qualities of a car depend on the amount it is allowed to roll on turns. When a car is designed to roll very much less than normal, and with a low c.g., so that the overturning couple on the tires is reduced, it may become necessary to put a strong understeering tendency into the rear axle control, to provide an adequate tail for the arrow." Tension type shackles are used to give a variable rate, but the static deflection at the rear is given as 5", equivalent to a ride rate at normal load of about 112 lbs/in., a ride frequency of 80 oscillations per minute.

The center of gravity of the Corvette is only 18" above the ground. This, in conjunction with a heavier stabilizer bar and stiffer, non-symmetrical rear springs contributes to greatly reduced roll angle when cornering.

Moment of inertia relative to the wheelbase of the car as whole is dependent on the distribution of the principal masses. A dumbell-like distribution, with weight concentrated at the ends, tends to give good riding qualities but a slow steering response. A sports car requires some compromise in this respect. In the Corvette the engine is

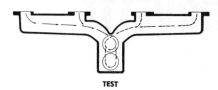

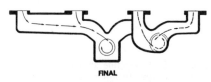

Dual exhaust adds to bhp, but thorough engineering resulted in over 4% better torque.

located 7 inches further back and 3 inches lower than a stock Chevrolet sedan. To avoid complexities, if we take the stock sedan's moment of inertia as 100%, the Corvette can be rated as 62%. Obviously much of this reduction in "dumbell" or flywheel effect comes about from the drastic reduction in body weight.

The net result of the suspension features just described more than fulfills the original design objective—smooth yet firm suspension. Again quoting Mr. Olley "A joggling ride is not acceptable, but a floating ride which appears to be divorced from the road is even more unacceptable. Excessive roll and vague handling characteristics will not do."

The question of weight distribution, fore and aft, should not be confused with mass distribution. A car could have its engine in the middle of the wheelbase and still have a 50/50 weight distribution. But its moment of inertia would be even less than the Corvette. Most engineers will agree with Mr. Olley that any deviation from 50/50 distribution should favor a slight nose-heaviness. There are many reasons for this, but

the most important one is that extra weight on the front wheels gives better directional stability at high speeds. The Corvette road tested this month (see page 10) weighed exactly 2890 lbs with full tank, spare tire, radio and heater. In this condition the fore and aft distribution was 1560 lbs front, 1330 lbs rear, or 54/46. With 320 lbs for driver and passenger added, the weights change by just over 2%, to 52/48. A mere 90 lbs of luggage over the rear axle changes the distribution another 2%, to 50/50. It is worth noting that every principal sports car manufacturer of note shows tendencies towards more weight forward. Allard's latest JR model is a complete reversal of previous policy with 57/43. Even Porsche is "reversing" their engine location from just behind to just ahead of the rear axle, on their latest type 550 model.

Having dealt with the various aspects of the chassis design and its effect on the all-important handling qualities, let us examine the rest of the Corvette. Many people like to point out that "Chevrolet hasn't changed their engine since 1937" (when four main bearings and a new stroke/bore ratio were adopted.) This is however, a compliment for it attests to the excellence of a design which, though perhaps not exciting or dramatic, has stood the test of time. Actually, very little of the original design is left. Today all Chevrolets have a larger bore and stroke, pressure lubrication to the rods, aluminum pistons, insert type bearings, an even heavier crankshaft, very large ports and valves. The accompanying power curve data compares the 108 bhp 1953 Chevrolet to the Corvette. The Corvette engine is not changed radically over the 1954 115 and 125 bhp stock engines. Compression ratio is slightly higher at 8.0 to 1. Adjustable type valve tappets run on a revised camshaft having .005" more lift than the production 125 bhp engine. The following table illustrates the moderate changes made in valve timing.

ENGINE	I.O.	I.C.	E.O.	E.C.
115 bhp	1.0	39.0	42.0	9.0
125 bhp	10.5	53.5	49.8	15
150 bhp	19.5	44.5	59.0	5.0

The corresponding valve spring pressures on the three engines are 160, 182 and 207 lbs. The camshaft timing gear is aluminum

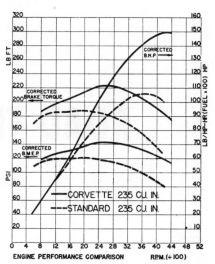

This shows the improved engine performance of the Corvette compared to the 1953 engine.

on the 150 bhp Corvette, to withstand the heavier loads and higher speeds.

There are three Carter side draft carburetors mounted on a cast aluminum manifold provided with suitable balancing passages. Automatic chokes were tried and abandoned because of choke valve flutter and fast idling. No exhaust heat is used or is necessary, but provision is made for extremely cold weather. The heat shield can be removed and the heat control valve spring reversed, if desired. The dual exhaust system (illustrated) is a special type to keep the gasses in the outlets always swirling in the same direction. This one feature added 8 to 10 ft/lbs of torque in the mid-speed range. On the mufflers, Mr. Olley says "A requirement in the minds of sports car enthusiasts is that the exhaust should have the right note. They don't agree what this is. Some prefer 'foo-blap' while others go for 'foo-gobble.' It is impossible to please them all. We hope we have achieved a desirable compromise."

The water pump is a special high-efficiency type running at .9 engine speed. Circulation rate is 27 gpm at 2000 rpm. The fan is 18 inches in diameter, not shrouded. Cooling tests show that the cooling system of the

(Continued on page 15)

The Corvette engine sits horizontal in order to get the drive-line over the X-member.

ROAD TESTING THE CORVETTE

is it really a sports car?

pulling the powerglide selector lever quickly from neutral to low range. Our acceleration times quoted were all made using drive range and normal starts. We experimented with "jerk" starts and found no better times from zero to any speed. The time for zero to 60 mph could be improved very slightly by placing the selector in low range. The car always starts in low anyway, but this procedure forces the transmission to stay permanently in first speed, where incidently a speedo reading of just over 70 is possible with the tachometer well past the last calibration mark of 5000 rpm. Strangely enough the forced low gear technique with a quick shift to drive at an honest 60 mph gave slightly slower times to 70 mph.

Probably no part of the Corvette specification is more controversial than its torque converter with 2 speed automatic transmission. Admittedly it will convert a lot of people to sports cars, who have no desire

Cylindrical tank supplements radiator top tank. Below this is the ignition shielding. Detail (below) shows side curtain vent.

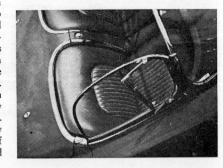

Ever since the Chevrolet Corvette was announced over a year ago, there has been much speculation over its competition performance potential. The die-hards, the pro-foreign advocates have been especially loud in their derision of the new car, maintaining that the Corvette is not a genuine dual-purpose sports car, but more of an effete high-speed touring type. Some have been more specific, claiming that nothing from Detroit could possibly be any good—least of all from Chevrolet.

This extreme attitude was not corrected by the Chevrolet Motor Division's ambiguous statement that their new sports car "is not intended to be used as a racing car". At this point it might be interesting to compare the attitude of Jaguar Cars, Ltd., at the time of the announcement of the XK-120. The Jaguar roadster was intended primarily as a high speed touring car and they were quite surprised to find it being raced so extensively in America.

Furthermore, some people seem to feel that no car based on standard family car components can be much of a sports car. Nothing could be further from the truth, as can readily be shown by mentioning such famous makes as Mercedes, Porsche, Alfa

Romeo, Siata, Lancia, Gordini, Talbot, and Jaguar—all of whom at one time or another built rather successful sports cars using a large proportion of major units from a mass-produced family type automobile.

So it is that the only fair approach to the Corvette must be on the basis of its all-around performance in comparison to other sports cars and completely ignoring the fact that it happens to stem from the world's largest producer of automobiles.

The Corvette makes a favorable impression immediately on the score of clean lines with a minimum of chrome trim. It looks like a sports car, a very modern one at that, and its wrap-around windshield alone indicates the trend of things to come. When first driving the car there is a tendency to keep reaching for the gear shift lever and the total absence of a clutch pedal is disconcerting. The accelerator, like Chevrolets from the year one, has a very stiff return spring which makes you feel like you are working hard to force the car to its limit. The initial surge is just a little sluggish and there is no problem with unnecessary (and fruitless) wheelspin, on dry pavement. Rubber burning starts can be made by grandstand drivers using the technique of turning the engine at 2000/2500 rpm and to develop driving skill. Admittedly it gives a tremendous performance through a well graduated series of infinitely variable ratios. It might even work fairly well in a road race—up to a point. That point is the serious question of safety when taking a fast turn at the ragged edge of tire adhesion. Suppose you are doing 45 mph at the time and need more power to pull-out of an incipient spin. A jab on the throttle, if done too energetically, will force a downshift to low and the ensuing jerk will certainly cause loss of control.

On the other hand the Corvette is supposed to bring the sports car to the mass market since its price is only a little more than a fully equipped Chevrolet convertible. As it is, only a very small percentage of sports car buyers race their cars and perhaps the automatic transmission will sell more cars. But the fact remains that we, as well as the great majority of sports car fans, would much prefer to have a traditional sports car stepped transmission. Four close ratios, all synchonized, would be perfect but even a standard Chevrolet 3 speed unit should be made available. This would give overall ratios of 10.44, 5.96 and 3.55, with corresponding speeds of 38, 67 and 112 mph at 5000 rpm.

The outstanding characteristic of the Corvette is probably its deceptive performance. Sports car enthusiasts who have ridden in or driven the car without benefit of stop watch seem to have been unimpressed with the performance. This is an injustice, as the figures shown in our data panel prove.

It should also be borne in mind that a "stick shift" Chevrolet sedan will accelerate to 60 mph in 2 seconds less time than a comparable year and model sedan equipped with Powerglide. Although it is doubtful whether a conventional transmission would make that much improvement in the Corvette the fact remains that the acceleration figures it gives are all the more remarkable. The Corvette will give any sports car of comparable power and weight a "real race" between 20 and 90 mph.

The timed stop speed runs were made with top and side curtains in place, as our experience has shown full enclosure invariably gives better results. In view of the fact that this car had only 500 miles on the odometer the best run of 107.1 mph is about right and a car with fully run-in engine should be able to top the 108 mph factory figure by a comfortable margin. After these runs there was a slight smell of burning paint, but 4700 rpm and a wide open throttle for about 12 miles on a new engine is asking quite a lot. During the strenuous acceleration checks the engine was taken up to over 5000 rpm at least 30 times. Once it went to an indicated 71 mph in low, an actual 68, without sign of valve bounce. That speed is 5500 rpm. At full throttle there is considerable air intake noise. The three sidedraft carburetors have small individual air cleaners which obviously have very little acoustical value.

The second most outstanding characteristic of the Corvette is its really good combination of riding and handling qualities. The ride is so good that few American car owners would notice much difference from their own cars. Yet there is a feeling of

(Continued on next page)

ROAD AND TRACK ROAD TEST NO. A-1-54

CHEVROLET CORVETTE ROADSTER

SPECIFICATIONS

Price, fob St. Louis	$3760
Wheelbase	102 in.
Tread, front	57 in.
rear	59 in.
Tire size	6.70x15
Curb weight	2890 lbs
distribution	53/47
Test weight	3210 lbs
Engine	6-cyl.
Valves	ohv
Bore & Stroke	3.56x3.94
Displacement	235.5 cu in.
	(3861 cc)
Compression ratio	8.00
Horsepower	150
peaking speed	4200
equivalent mph	94.6
Torque, ft/lbs	223
peaking speed	2400
equivalent mph	54
Mph per 1000 rpm	22.5
Mph at 2500 fpm	
piston speed	86
Gear ratios (overall)	
Low + converter	13.57
Low	6.46
Drive	3.55
R&T perf. factor	63.2

PERFORMANCE

Top speed (avg.)	106.4
fastest one way	107.1

Max. speeds in gears— Powerglide transmission in drive-range gives automatic up-shift from low to high at 58 mph under wide open throttle. Low range can be used up to 68 mph (5500 rpm).

Mileage .... 16/20 mpg

ACCELERATION

0-30 mph	3.7 secs
0-40 mph	5.3 secs
0-50 mph	7.7 secs
0-60 mph	11.0 secs
0-70 mph	14.8 secs
0-80 mph	19.5 secs
Standing start ¼ mile—	
average	18.0 secs
best	17.9 secs

TAPLEY READINGS

590 lbs/ton at 32 mph using full throttle which holds the transmission in low range. Tapley readings in high gear were not attempted because of the torque converter.

COASTING

(wind and rolling resistance)

90 lbs/ton	at	60 mph
45 lbs/ton	at	30 mph
30 lbs/ton	at	10 mph

SPEEDO ERROR

Indicated	actual
10	10.9
20	19.8
30	29.3
40	38.5
50	48.1
60	57.9
70	67.2
80	77.0

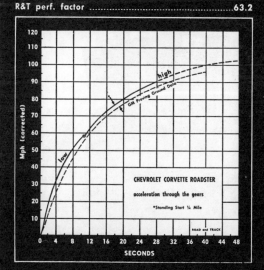

CHEVROLET CORVETTE ROADSTER

acceleration through the gears

*Standing Start ¼ Mile

ROAD and TRACK

firmness about the car, and none of the easy slow motion effect of our large heavy sedans. The biggest surprise is the low roll angle—actually less than two of the most popular imported sports cars. The Corvette corners flat like a genuine sports car should.

About 100 miles of highway driving was done during a heavy downpour. Speeds of 90 mph were attained on the wet roads with no feeling of insecurity. On dry surfaces

there is just the right amount of understeer, and caster action is pleasantly moderate. The steering is in fact very light, and with 3.7 turns lock to lock it could be faster with considerable benefit. There was more sensitivity to gusty cross winds than we liked and high speed four wheel drifts required a certain amount of dexterity that is not usually associated with a sports car. The road test car was equipped with prem-

ium grade tires (inflated to 30 psi). A short run in another Corvette with regular grade tires, having a wider flatter tread design, convinced us that the higher priced tires involve a definite loss in handling qualities.

The "unusual" California weather also provided an opportunity for a special brake test. Bendix duo-servo brakes are often extremely sensitive to water, but there were no such faults on this car. Judicious braking gave easily controlled deceleration in the wet, with no sign of grab or pull. It was also during this interval that we recorded the best fuel mileage of 20 mpg—cruising at a steady 55/60 mph. Consumption dropped to 16 mpg during the performance checks, a figure which is far lower than even the worst traffic driving would give.

Corvette top is neat, fairly easy to set-up and did not leak in a heavy rain storm.

Although being deliberately whipped around this turn, the Corvette leans very little. Trunk space (below) is large because gas tank location allows low mounting for spare tire.

During the rain storm, the top and very practical side curtains were completely effective, but the leading edge of the door opening leaked. The slight drip was however useful in putting out cigarettes. Aside from the leak, the fiberglass body showed absolutely no fault. There is no drumming or tendency to rattle and the general quality level of construction and finish is excellent. The 53° windshield gives good protection and visibility with the top either up or down. Interior roominess is unusually good for a two-seater sports car and there is ample room for the feet—a welcome relief after the cramped pedal space found on so many two-seaters. Adequacy of the bumpers may be criticized as witness the following note found in the Operations Manual—it reads, "Using the Corvette to push or pull other vehicles is not recommended."

Frankly, we liked the Corvette very much. It may not be suitable for road racing competition, as it comes from the factory—but very few sports cars are today. For those who want to compete with an American sports car, it should be easy enough to strip the car down to a better weight. There is even a new 261 cu in. (4278 cc) Chevrolet truck block to boost the engine size closer to the top limit for Class C (3000 to 5000 cc) and all the Corvette parts will fit, except pistons. And if you want a really competitive car the 4950 cc GMC-302 can be installed, offering a conservative 200 bhp on gasoline. You can even install a 4-speed Chevrolet truck transmission but the gear ratios are unsuitable. However, the Cyclone adapter enables use of the close ratio Lincoln-Zephyr transmission with either Chevrolet or GMC engines. Kelsey-Hayes wire wheels and aluminum muffs bonded to the brake drums should take care of the brake cooling. Its only a question of time before some or all of these modifications will be made by someone.

Finally, we would like to say a word in complimenting the Harry Mann Chevrolet Company in Los Angeles. Both Pete Mann and Frank Milne were extremely cooperative in arranging for us to test the Corvette. They even provided us with a 1954 Chevrolet sedan to transport the necessary road test equipment.

Chevrolet may have committed some errors in presenting and merchandising a sports car, but the people we met at this Chevrolet dealership were refreshingly alive to the special needs and desires of sports car enthusiasts. ●

Road Test: **The Corvette V8**

it may be "loaded for bear" but . . .

photography: **Ralph Poole**

Fiberglass top is not a factory option.

L AST YEAR, when we road tested the Corvette 6, the question was raised, "Is it really a sports car," Results to date are fairly conclusive, for the occasional rare appearance of a Corvette in competition has not been marked by any major upset.

For 1955 there are no changes in the car's specifications except that the new Chevrolet V-8 engine, tuned to 195 bhp, is optional, and the output of the 6 has been upped to 155 bhp (5 more). The V-8 powered version gives truly startling performance, as might be expected, but the transmission and brake deficiencies still will not satisfy the demands of either competition or of the true sports car enthusiast, no matter how loyal he may be to American engineering know-how. Furthermore, the theme "For Experts Only" can be applied only to those new to the sport, since the Corvette demands no unusual driver skill or special techniques. There is a rumor that a 3-speed manual transmission will be made available, but no cars have been built with this equipment as yet.

Nevertheless, the fact remains that there is a small market for an open roadster which the lady of the house can use as smart personal transport without the necessity of learning how to manipulate one of those "funny old-fashioned gear shift levers." Chevrolet's Corvette fulfills the above need admirably, and if sales have been meagre it can only be blamed on the fact that a "female" personal car should also cater to the comfort demands of the sex and provide the usual assortment of power-operated gadgets (including side windows), three tone color schemes and frilly curtains.

The amazing thing about the Corvette is that it comes so close to being a really interesting, worth-while and genuine sports car—yet misses the mark almost entirely. Last June we said "The outstanding characteristic of the Corvette is probably its deceptive performance." Quite naturally the more powerful V-8 gives vastly improved performance as the following comparison table will show.

	1954 Six	1955 V-8	
Top speed (best)	107.1	119.1	mph
0-30 mph	3.7	3.2	secs
0-60 mph	11.0	8.7	secs
0-80 mph	19.5	14.4	secs
0-100 mph	40.0	24.7	secs
SS¼ (avg.)	18.0	16.5	secs
Odometer	500	1450	miles

Despite the improvement in performance, economy has not been sacrificed; on the contrary the new low-friction engine yielded 2 to 3 miles more per gallon than last year's test of the 150 bhp 6 gave. The new V-8 engine is, incidentally, a stock passenger car unit with the four barrel "power-pack" carburetor. The only change is a special camshaft which alone accounts for the 15 extra horses. (The horsepower peaking speed and torque figures are our estimates, no official data being available). Reliability should be up to family sedan standards, and as a matter of interest the theoretical cruising speed, even with a 3.55 axle ratio, is nearly equal to the actual top speed—thanks to the shortest stroke in the industry (American). The new engine is very smooth and quiet, particularly on full throttle acceleration, whereas the 6 sounds a little on the "cobby" side under the same treatment. At idle there is some tappet noise, explained perhaps by settings slightly towards the high-limit, in anticipation of the high-speed runs. The 8 does idle much more quietly than the 6, however.

Last year we also mentioned that "The second most outstanding characteristic of the Corvette is its really good combination of riding and handling qualities." We added, "It may not be suitable for road-racing

A minor change in the script lettering on the side identifies the 1955 Corvette V-8.

Corvette cockpit is spacious and luxurious.

as it comes from the factory—but . . . it should be easy enough to strip the car down to a better weight". Watching a Corvette in an airport race coming into a corner with fast company, we have observed that the brakes show up poorly, but the actual cornering is done just as fast, flat and comfortably as several imported sports cars we could name. Coming out of the corner, the Corvette 6 seems to accelerate on a par with an average 2-litre production sports car (as it should on the basis of performance tests) but not nearly as well as cars in its own engine-size class having four forward speeds and approximately the same power-to-weight ratio. The new V-8 powered car still will need "brakes", but it handles very well and should now accelerate more on a par with machines of its class. Worth mentioning is the fact that the brakes are more than adequate for ordinary usage, with 158 sq. in. of lining area for only 2900 lbs of machine (109 sq. in/ton).

Chevrolet says the Corvette has "quick" 16:1 steering, but 3¾ turns, lock to lock, is not quick-steering "for experts", nor fast enough for a sports car which merits our whole-hearted and unreserved recommendation. The steering is actually very light at all times and could easily be reduced to 3.0 turns lock to lock and still be parkable, especially if power-steeriig were available for the frailer type of driver.

Riding qualities are excellent and directional stability at high speeds is near-perfect. One does wonder why the much advertised ball-joint front suspension of the

(Continued on page 15)

The V8 engine is 41 lbs lighter than the 6.

14

ROAD & TRACK ROAD TEST NO. A-4-55
CHEVROLET CORVETTE V-8

SPECIFICATIONS

List price	$2901
Wheelbase	102 in.
Tread, front	57 in.
rear	59 in.
Tire size	6.70-15
Curb weight	2880
distribution	52/48
Test weight	3200
Engine	V-8
Valves	pohv
Bore & stroke	3.75 x 3.0 in.
Displacement	265 cu in. (4344 cc)
Compression ratio	8.00
Horsepower	195
peaking speed	4600
equivalent mph	103
Torque, ft/lbs.	260
peaking speed	2800
equivalent mph	63
Mph per 1000 rpm	22.5
Mph at 2500 fpm	112.5
Gear ratios (overall)	
Drive	3.55
Low	6.46
Low + converter	13.57
R & T high gear performance factor	72.5

PERFORMANCE

Top speed (avg.)	116.9
best run	119.1
Max. speeds in gears—	
low (5500 rpm)	68

N.B.—Powerglide transmission normally shifts from low to high at 62 mph, under wide-open throttle.

Mileage	18/22.5 mpg

ACCELERATION

0-30	3.2 secs.
0-40	4.4 secs.
0-50	6.4 secs.
0-60	8.7 secs.
0-70	11.3 secs.
0-80	14.4 secs.
0-90	18.9 secs.
0-100	24.7 secs.
Standing ¼ mile—	
average	16.5 secs.
best	16.3 secs.

TAPLEY READINGS

Gear	Lbs/ton	Mph	Grade
Low	600	35	32%
High	380	50	19%

Total drag at 60 mph, 152 lbs.

SPEEDO ERROR

Indicated	Actual
10	11.0
20	20.4
30	29.7
40	39.5
50	48.6
60	58.0
70	67.3
80	77.0
90	86.1
100	97.1
118	119.1

CHEVROLET CORVETTE V-8
acceleration through the gears

ROAD and TRACK

(Graph: Mph (corrected) vs SECONDS, showing acceleration curve for low and drive gears)

Corvette . . .
(Continued from page 14)

1955 passenger cars was not applied to the sports car to give the reduced brake dive, easier maintenance and longer life of the ball joints. However, the ball type i.f.s. would have meant an extensive re-design in order to incorporate it—for 1955.

This year we were not "favored" by inclement weather for the purposes of testing, but judging from the number of letters we have received, the average Corvette of 1954 was not notable for its freedom from water leaks during a downpour. There is no reason to believe that the 1955 model will be any better. Our test car was equipped with a fiberglass top which is not a factory-approved extra (made by Plasticon, $225) but it smooths up the car's appearance and is almost instantly removable with the aid of four toggle type clamps.

Externally the Corvette scores heavily because it has well-executed sports car styling. There has been ample criticism of the fiberglass body material employed, but most if not all of this can be traced to the lack of adequate bumpers—a strange complaint to be applied to an American car. The weight-saving advantages of a fiberglass body should appeal to sports car people, and the new V-8 engine weighs 41 lbs less than the 6. However, our test car weighed only 10 lbs less than last year's car, since two tops were carried.

The instrument panel layout of the Corvette commits a cardinal sin by using a very small tachometer and placing it in the center of the panel rather than alongside the speedometer. Real oil-pressure and ammeter gages are a saving grace in this day of warning indicators.

The seats are beautifully done and very comfortable but no provision has been made for seat belts. A large central transmission cover is expected in a sports car, and the Corvette's is typical. Knowing the compactness of a Powerglide transmission unit does raise a question as to why this cover had to be so large, even though there is ample pedal room.

Finally, to divert an avalanche of Technical Correspondence, let us mention that the Chevrolet V-8 engine can be (and has been) installed in last year's Corvette. For that matter, so can many other V-8 engines including Cadillac, Buick and Oldsmobile. However, even the lightest of these 3 (the Buick) adds over 100 lbs to the front end weight.

Touring in a sports car was never like this.

CORVETTE
(Continued from page 9)

Corvette (pressurized at 4 psi) is far above normal passenger car standards.

The ignition system is 6 volt, with a special coil, condenser and distributor cam. Voltage reserve is ample for speeds well above 5000 rpm. Standard spark plugs are 14mm AC 44-5, but the AC 43-5 is recommended for continuous high speeds.

As a result of these engine modifications the maximum output has been raised to 150 bhp at 4200 rpm. More power has been developed experimentally, but only at a higher peaking speed, and accompanied by a serious loss in torque. As it is, the bhp gain is 20%, yet torque has also gone up by 11.5%. The net result is a vastly better acceleration curve, and a smooth idle at 475 rpm.

For reasons of wider appeal, a modified Powerglide transmission is used on the Corvette. Mr. Olley says on this: "The use of an automatic transmission has been criticized by those who believe that sports car enthusiasts want nothing but a four speed crash shift. The answer is that the typical sports car enthusiast, like the 'average man', or the square root of minus one, is an imaginary quantity. Also, as the sports car appeals to a wider and wider section of the public, the center of gravity of this theoretical individual is shifting from the austerity of the pioneer towards the luxury of modern ideas . . . there is no need to apologize for the performance of this car with its automatic transmission." That statement, from Chevrolet, should get a rise from 100,000 *Road & Track* readers!

The rear axle is essentially stock, but the housing and pinion gear are special to provide an oil seal and universal joint flange. The open drive shaft is only 36 inches long and check straps are used to prevent too great an angle at full rebound. The spring pads are also different from the stock rubber bushed pin used with the torque tube.

The first Corvette shown at G.M.'s New York Show in January 1953 had a fiberglass body, as is well known. What isn't so well known is that the 1954 production schedule of 10,000 units were to have steel bodies made from Kirksite dies. But, the demand for immediate delivery was so great that it was decided to build 300 Corvettes in 1953, with fiberglass bodies. As experience was gained with the new material, so also did confidence increase. The result was the decision to build the entire contemplated 1954 production in fiberglass, and the Kirksite dies were never cast.

Chevrolet sums up the experiment this way. "What we get for all this is a very usable body, somewhat expensive, costing a little less than a dollar a pound, but of light weight, able to stand up to abuse, which will not rust, will not crumble in collision, will take a paint finish, and is relatively free from drumming noise."

Finally it is worth noting that it is amazing to find a great mass production organization willing and able to step out of its normal role of producing over 500 vehicles an hour, to make 500 specialized vehicles in two weeks. The Corvette is more than just a new sports car. It is all of that, but perhaps more important it heralds an entirely new approach, offers new hope, for the individualist. ●

Corvette

The Corvette corners as a sports car should, fast and flat.

CONTINUED FROM PAGE 19

larger air cleaner-silencer. The interior treatment is impressive but the new winding windows and the power operated top have forced some curtailment in elbow and leg room. The top, incidentally, is only semi-automatic for it must be released and partially collapsed before pressing the fold button.

Since there is considerable confusion over the price of the Corvette, the essential data is tabulated herewith. The list or base price is f.o.b., St. Louis and includes a 210 bhp engine with one four-barrel carburetor, the 3-speed transmission and a manually operated soft top.

Specifications

List price, f.o.b.	$3120.00
Dual carburetors	172.20
Powerglide	188.50
Power windows	64.60
Radio & heater	322.55
Power top	107.60
Hard top	215.20
in place of soft top	N.C.
3.27 axle ratio	N.C.

Other options, such as white wall tires, windshield washer, power steering, etc., are available. An interesting item for the enthusiast is a special high lift camshaft, "recommended for racing purposes only." The extra charge is $188.30 and it gives the dual carburetor engine an output of 240 bhp. Also optional at unstated prices are segmented-metallic brake linings, magnesium wheels and Firestone SS-170 tires.

Again we are indebted to the Harry Mann Chevrolet Company of Los Angeles, this being the third Corvette they have supplied us for testing purposes. The stick-shift car is the property of Ralph Petersen whose enthusiasm has no limit. This is his third Corvette and when the new high lift camshaft and 4.1 gears are installed, he will enter the car in competition. ●

The car can be ordered with either a soft or a hard top at no difference in price.

The new, optional hard-top version of the Corvette.

1956 CORVETTE
100 mph in 2nd!

With .85 hp/cu in, the Corvette develops more power for its size than any other American engine. Metal coverings over the ignition system are for eliminating radio interference with non-metallic body.

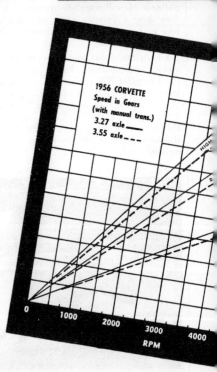

1956 CORVETTE
Speed in Gears
(with manual trans.)
3.27 axle ———
3.55 axle - - -

0   1000   2000   3000   4000

RPM

Rear view shows changes in fenders, side panels, tail lights and "bumpers".

AMERICA's only production sports car (by most definitions) has been restyled for 1956, but the most interesting part of the new Corvette is in the technical specifications department.

The light-weight V-8 engine has the same displacement as last year's Corvette, but the output has been increased from 195 bhp at 4600 rpm to an amazing 225 at 5200. Two, 4-barrel carburetors, a new compression ratio of 9.25-to-1 and a revised camshaft are the principal contributors to the higher output. Also of interest is the change in torque curve characteristic, from 260 at 2800 to 270 at 3600.

Performance-wise the new car should prove substantially better than our last year's test on the V-8 model (R & T for July, 1955). Not only are horsepower and torque increased but a manual-shift transmission is available, at last. At first glance only 3 forward speeds may appear as a disappointment, but the gear ratios are very close, 2nd gear, for example, being closer to high than most 4-speed sports cars. Low, or 1st gear, is approximately similar to 2nd gear in an imported sports car—exploiting the advantage of ample cubic inches. The power-glide automatic transmission is still available, and in addition there is a choice of two axle ratios; 3.55 (as before) or 3.27, a new option. At the engine's peaking speed of 5200 rpm the 3.55 axle gives 117 mph, the 3.27 axle gives 127 mph, with no allowance for tire expansion. True timed top speeds should be slightly above these figures.

Body material remains fiberglass, now moulded completely in matched metal dies. The front fenders are revised to accommodate normal type headlights while the jet-spinners on the rear fenders have been eliminated. The doors are equipped with wind-up windows and power operation is available at extra cost. The windshield is new and the standard cloth top has more bows and is power operated. Although the exhaust outlets have been relocated to eliminate trouble with soot accumulation, there are still no real bumpers and the grille is unchanged. An optional hard-top with rear quarter windows will be available, exactly as shown on R & T's cover for June, 1954. The new spring-spoked steering wheel looks as if it came from Italy but makes the instrument panel layout, which is unchanged, look a little dizzy. The floor mounted "stick-shift" is nicely done and has an ash tray alongside it on the tunnel.

The new clutch is 10" in diameter and uses a coil spring type pressure plate assembly. The differential unit is also new, as redesigned for the 1956 passenger car. An interesting option is four-ply, high-speed nylon racing tires, which certainly will be necessary if the top speed potential of this car is ever used. The front suspension is unchanged since the Corvette frame is designed to use the 1951/1954 Chevrolet suspension with its "integral" front cross member. Brakes too, remain as before, 11" Bendix.

The 1956 Corvette should prove to be a very interesting car for enthusiasts who have asked for an American dual-purpose sports car. It can run as a production car in Class C events (3 to 5 litres) and should liven up proceedings in this category considerably. ●

1956 Corvette Specifications

Wheelbase, in.	102
Tread, front	56.7
rear	58.8
Tire size	6.70-15
Curb weight (est.)	2880
Engine	V-8
Valves	pohv
Bore & stroke, in.	3.75 x 3.0
Displacement, cu. in.	265
cc	4344
Compression ratio	9.25
Horsepower	225
peaking speed	5200
Torque, ft/lbs.	270
peaking speed	3600

Chart at the left gives the mph at any speed in any gear with the new close-ratio, 3-speed transmission. Based on 750 tire revs per mile.

Road Test: Two Corvettes

0-60 in 7.3 seconds
0-100 in 20.7 seconds

VERY FEW NEW sports cars have aroused as much interest or created as much controversy as has the 1956 Corvette. The first real test of the new Corvette came at Pebble Beach (see page 22) and its performance there in a race restricted to over 1500 cc production cars, was the surprise of the day (it finished 2nd). On paper, and according to the data in this road test, the new Corvette is the best performing production car in class C today; at least until the S.C.C.A. rules that the D-type Jaguar is a "production" car by virtue of just over 100 having been built.

In June of 1954, we published a road test on the Corvette 6, and the 195 bhp V-8 version was tested in July, 1955. This report covers two 1956 models, one with the new stick-shift, the other with Powerglide. The car with the automatic transmission was tested primarily for comparison purposes against the two earlier tests, since they had a similar transmission. All four cars, fortunately, had the same axle ratio of 3.55-to-1. A comparison tabulation is very interesting and is as follows:

Corvette Comparison

Year	1954	1955	1956	1956
Trans.	P.G.	P.G.	P.G.	Stick
Bhp	150	195	225	225
Curb wt.	2890	2880	3080	2980
Test wt.	3210	3200	3410	3330
Top speed	104.4	116.9	121.3	129.1
0-60	11.0	8.7	8.9	7.3
0-80	19.5	14.4	14.4	12.4
0-100	41.0	24.7	24.0	20.7
SS1/4	18.0	16.5	16.5	15.8

photography: Poole

This table shows the tremendous improvement in performance that three years of development has accomplished. It shows very well the advantage of the new stick-shift transmission, but it must be mentioned that the 1956 Powerglide car had only 600 miles on the odometer which may have reduced the top speed somewhat. It also shows that the new Powerglide model is approximately 200 lbs. heavier than the '55 V-8, and 100 lbs. heavier than the '56 with stick-shift. (Both cars had R & H, power top, etc.)

The first two high speed runs with the stick-shift car recorded an average speed of exactly 125.0 mph. The speedometer held steady at 124 mph, but the tachometer seemed to lag at about 5200 rpm. Four more runs, starting with a much longer "wind-up" produced a very wavering speedometer reading, which varied all the way from 130 to 140 mph. The tachometer also behaved somewhat erratically, but read about 5700 rpm on the two best runs, which averaged 129.1 mph. This car had nearly 3000 miles on the odometer and the speedometer was almost perfect in its accuracy up to 125.

During the acceleration tests, both cars displayed fairly pronounced "flatness" of carburetion on the take-off. This appears to be a characteristic of the two four-barrel carburetors and possibly accounts for the change to a single four-barrel carburetor as standard equipment (210 bhp). The Powerglide car gets away from a standstill better than the stick-shift, on the initial "jump." Its time to 60 mph was .2 seconds slower than last year's car, because of the added weight and carburetion fault. However, from 60 mph upwards, the extra 30 bhp begins to tell and it was .7 seconds better to 100 mph and 4.4 mph higher in top speed than the 1955 test. Despite the slightly slower start, the '56 Powerglide car recorded the standing start 1/4 mile in 16.5 seconds (average), identical to last year.

The stick-shift model takes off unimpressively too, primarily because of the very "high" first gear (7.81 overall), as compared to the 11.9 starting ratio on the Powerglide. However, at 25 mph in first gear, the power really comes on like a blast and 60 mph can be touched in very close to 7 seconds dead by over-revving somewhat. The new floor shift works well though just a little heavy when trying hard. Second gear gives tremendous acceleration all the way to 100 mph if desired. Even third, or high gear, pulls well as witness the Tapley reading of 350/ton —better than most sport cars in their next-

The fiberglass body is continued with new front fenders and indented side panels.

Instrument panel is unchanged but the speedometer error on both cars was practically nil.

to-high gear. An optional axle ratio of 3.27 is available and while it might add 4 or 5 miles to the top speed, we feel that the 3.55 ratio in our test car is a good choice because of the "high" low gear.

The new close-ratio transmission has some unusual features. In the first place, it is somewhat noisy and we are told that the gears are all "straight-cut" (not helical), which seems plausible. Secondly, although first gear is not synchronized, it can be engaged at 50 or 60 mph without double-clutching. The technique requires only that the engine speed be brought up during the down-shift, *while the clutch is depressed.* Such shifts are almost fool-proof and require very little finesse, but the same procedure at 25 or 30 mph doesn't work so reliably and in this case the usual double-clutching process is safer. The advantage of being able to use first gear, while slowing down for a corner, (in a race) is of course considerable and is a feature of the new Corvette which was certainly not expected by us.

The general handling qualities and cornering ability of the Corvette remain "good to excellent" as compared to other dual-purpose sports cars. We did notice, for the first time, a certain amount of body and cowl shake at over 100 mph, which may be due to the very high speeds the new car attains so readily. The more powerful engine is smooth all the way to nearly 6000 rpm, but it did seem a trifle noisier, under full throttle, than last year's car, which had a much

CONTINUED ON PAGE 15

Two 4-barrel carburetors provide a carburetor for each cylinder.

ROAD & TRACK ROAD TEST NO. A-3-56

TWO CHEVROLET CORVETTES

SPECIFICATIONS

List price (stick-shift)	$3120
Wheelbase	102 in.
Tread, front	56.7 in.
rear	58.8 in.
Tire size	6.70-15
Curb weight	2980 lbs.
distribution	51/49
Test weight	3330 lbs.
Engine	V-8
Valves	pushrod ohv
Bore & stroke	3.75 x 3.0 in.
Displacement	265 cu. in. (4344cc)
Compression ratio	9.25
Horsepower	225
peaking speed	5200
equivalent mph	117
Torque, ft/lbs	270
peaking speed	3600
equivalent mph	81
Mph per 1000 rpm	22.5
Mph at 2500 fpm	112.5
Gear ratios (overall)	
3rd	3.55
2nd	4.65
1st	7.81
R & T high gear performance factor	73.0

PERFORMANCE

Top speed (avg.)	129.1
best run	130.2
powerglide	121.3
Max. speed in gears—	
2nd (5800)	100
1st (5900)	60
Shift points from—	
2nd (5500)	95
1st (4900)	50
Mileage	13/16 mpg

ACCELERATION

0-30 mph	2.7 secs.
0-40 mph	3.9 secs.
0-50 mph	5.3 secs.
0-60 mph	7.3 secs.
0-70 mph	9.8 secs.
0-80 mph	12.4 secs.
0-90 mph	16.0 secs.
0-100 mph	20.7 secs.
Standing ¼ mile—	
average	15.8 secs.
best	15.7 secs.

TAPLEY READINGS

Gear	Lbs/ton	Mph	Mph/sec
1st	off scale		
2nd	460	62	4.6
3rd	350	70	3.5

Total drag at 60 mph, 132 lbs.

SPEEDO ERROR

Indicated	Actual
30	30.2
40	40.1
50	50.0
60	60.0
70	69.5
80	79.6
90	90.0
100	100.0
125	126.2

CHEVROLET CORVETTE COMPARISON
acceleration thru the gears
1956 with stick shift ———
1956 with power glide - - - - -

ROAD and TRACK

FROM COVENTRY

THE JAGUAR SS

FROM DETROIT

THE CORVETTE SS

ALTHOUGH sports cars still do, and probably always will, constitute only a small market for the volume-mad producers of automobiles, a firm such as Jaguar builds nearly half of its machines in sports car form.

Accordingly, when Jaguar announces a new model it *is* news, and the new XK-SS is going to be something of a sensation. Any enthusiast will instantly recognize that the "SS" is basically a D-type, and the first question will concern price. The port-of-entry-duty-paid price is $5600, a remarkable figure for a machine of this caliber which will be built only in limited quantities.

The new car fully meets the rules and regulations of the Sports Car Club of America as to the definition of a "production-sports" model, and is in fact an excellent example of a genuine dual-purpose machine—in marked contrast to a trend (in some areas) towards effeminate, super-luxurious two-seaters.

Nevertheless, the interior comfort features of the SS are truly typical of past and present XK-Jaguars, and the top and side-curtains are carefully executed to complement the "daily-driving" side of its dual nature.

As for the competition side of the picture, a power to weight ratio of under 9 lbs/hp (with driver) takes care of that and one can only hope that other manufacturers will now be encouraged if not forced into offering something competitive.

The specification panel which appears here must be regarded as provisional, since full details are not yet available. However, we understand that carburetion by 3-dual Webers and multiple-spot brakes by Dunlop are included. ●

BRIEF SPECIFICATIONS

Wheelbase, in.	90.6
Tread, front	50.0
rear	48.0
Tire size	6.50-16
Dry weight	2040
Engine	6-dohc
Bore & stroke	3.27x4.17
Displacement, cu in.	210
cu cm.	3442
Compression ratio	9.0
Horsepower	262
peaking speed	6000
Axle Ratio, std.	3.54
optional	2.92, 3.31, 3.77
Transmission ratios — 4th	direct
3rd	1.28
2nd	1.64
1st	2.14
Performance factors:	
lbs/bhp (300 lbs. added)	8.94
cu ft/ton mile	128.0
engine revs/mi.	2460
piston travel, ft/mi.	1720
mph @ 2500 fpm	87.9
Estimated performance in full touring trim:	
top speed, mph	146
0-60 mph, secs.	5.5
0-100 mph	13.5
ss ¼ mile	14.1

ANNOUNCED at the annual meeting of the S.C.C.A. in Detroit, the Super-Sport version of the 1957 Corvette is the car which advocates of Detroit "know-how" have been waiting for. It also shows that Detroit "can do" if they so desire.

The SS model is essentially a competition version of the 1957 Corvette in which various options have been consolidated. The basic list includes the 283 hp fuel injection engine, a single-plate clutch, the long awaited 4-speed transmission, a limited-slip differential, metallic brake linings, finned brake drums, special (heavy-duty) springs, shock absorbers and stabilizer, and air ducts for the brakes.

The car shown has a special cowl with double-bubble windscreens and no top or windows, but it was built for "show" and apparently will not be available to the public.

The new gearbox is surpisingly compact with all four forward speeds synchronized. First and 3rd gears have ratios identical to the 3-speed box's 1st and 2nd. The main housing is cast iron and reverse gear is situated in the rear bearing retainer cover.

The 283 hp engine package includes the mysterious "Duntov" camshaft. This cam is not particularly unusual; valve lift is .3938" (int.) and .3997" (exh.), but duration is quite long at 287° for both intake and exhaust. It has proper ramps for adjustable tappets and comes with special 238 lb. dual springs.

The 1957 cars will incorporate some minor changes in the front suspension and steering and the rear suspension has been re-arranged for better rear end adhesion. Chevrolet proclaims the Corvette as "America's only true sports car," and their racing record for 1956 bears this out. ●

BRIEF SPECIFICATIONS

Wheelbase, in.	102
Tread, front	57.0
rear	59.0
Tire size	6.70-15
Dry weight	2737
Engine	V8, ohv
Bore & stroke	3.875x3.0
Displacement, cu in.	283
cu cm.	4639
Compression ratio	10.5
Horsepower	283
peaking speed	6200
Axle Ratio, std.	3.70
optional	3.55, 4.11, 4.55
Transmission ratios—4th	direct
3rd	1.31
2nd	1.66
1st	2.20
Performance factors:	
lbs/bhp (300 lbs. added)	10.71
cu ft/ton mile	149.5
engine revs/mi.	2770
piston travel, ft/mi.	1385
mph @ 2500 fpm	108.5
Estimated performance in full touring trim:	
top speed, mph	138
0-60 mph, secs.	6.3
0-100 mph	17.5
ss ¼ mile	15.2

PEBBLE BEACH

Production Car Race, Over 1500 c.c.

1st in Class C – CORVETTE
2nd Over-all – CORVETTE

FINISH

The 1956 Corvette is proving — in open competition — that it is America's only genuine production sports car. Extremely stable, with outrigger leaf springs at the rear, 16 to 1 steering ratio and weight distribution close to 50-50, it has demonstrated its ability to corner on equal terms with the best European production sports cars. Of particular interest to the competition driver are the close-ratio manual gearbox, with 2.2 to 1 low gear and 1.31 to 1 second gear; the two rear axle ratios of 3.70 to 1 or 3.27 to 1; the power-to-weight advantage of the Corvette's glass fiber body and the special racing brake lining, now available at Chevrolet dealerships.

Mainspring of the Corvette's per-formance is, of course, the fantastically efficient 4.3-litre V8 Chevrolet engine. Holder of the Pikes Peak stock car record, heart of the Corvette that set a two-way American mark of 150 m.p.h. at Daytona Beach last January, powerplant of the NASCAR stock car Short Track champion, this short-stroke V8 is capable of turning well over 5000 r.p.m. The Corvette makes it available in two versions: 210 h.p. with the single four-barrel carburetor, and 225 h.p. with dual four-barrel carburetors. A special high-performance cam is optional at extra cost.

For the driver who requires a superlative touring car rather than a competition machine the Corvette is available with such extra-cost lux-uries as removable hardtop, power-operated cloth top and window lifters, plus the Powerglide automatic transmission. And, for either choice, there is the assurance of reasonably priced service and parts at any of nearly 7500 Chevrolet dealerships.

If you are considering a sports car, don't fail to sample the 1956 Corvette. It is the surprise car of the year, as your Chevrolet dealer will be delighted to demonstrate.

CHEVROLET

SS MARK II

Flint's little Corvette is growing up

In the new SS, rear springs, shock absorbers and spring bumpers, like those at the front, are of unit construction and are assembled before installation at each wheel. Four arms control movement at the rear wheels, linking the de Dion tube with the frame. The arms are ball jointed at the tube and mounted in rubber to the frame.

Hinged at the front (just forward of the cowl) and at the rear, the magnesium alloy body looks like a cross between a D Jaguar and a showroom model Corvette. The curious head rest housing encloses a roll bar. The windscreen is plastic.

GENERAL SPECIFICATIONS

Wheelbase, in.	92
Tread, front and rear	51.5
Tires, front	Super Sports, 6.50/6.70-15, 6 ply
rear	Super Sports, 7.10/7.60-15, 6 ply
Dry weight	1850
Engine	V-8, ohv, fuel injection
Bore & stroke	3.875x3.0
Displacement, cu. in.	283
cu cm.	4639

AT THE OUTSET, we'll say that we weren't pulling your leg last month with our story on the Corvette SS, and that we are doing so this time with the title on this page. Both last month's double-bubble car and this are one-off versions; this one might be described as even more so.

Zora Arkus-Duntov is responsible for the SS. Last October, he was assigned an engineering staff and perhaps the biggest carte blanche ever issued at General Motors. E. R. Cole, Chevrolet's general manager, says of the SS: "It is a study of new ideas to determine whether they might eventually be refined and offered in regular passenger cars. Instead of substituting these features in test cars on a piecemeal basis, we hand-built a car around them that will provide concentrated results. Testing over tough race courses will serve to quickly furnish comparative engineering data that, under ordinary circumstances, will require long periods of research." We have no comment on that; let's get to the car itself.

Basically the stock 283 cubic inch engine, the SS version has aluminum cylinder heads, clutch housing, water pump and radiator core. The oil pan is magnesium. Engine weight is about 1.5 pounds per horsepower. The fuel injection is economically described as "experimental"; the radiator grille lets unheated air into a plenum chamber. Valve tappets are naturally solid, but the clutch is hydraulic.

There is no fan. Engine oil is cooled at the bottom of the radiator, and ducts carry air over the engine and front brakes. Two separate systems give individual power assistance to front and rear brakes (the rear ones are mounted inboard). A mercury switch can be adjusted to keep the rear wheels from locking in fast deceleration. Two leading shoes are used. Drums are cast iron with a cast aluminum outer rim, for rigidity plus heat dissipation.

An aluminum alloy case keeps weight of the entire 4-speed, fully synchronized transmission to 65 pounds. A short propeller shaft connects to the differential, mounted rigidly on the frame.

A recirculating ball type steering gear has its linkage ahead of the front wheel spindles. Overall ratio is 12:1, and the handsome wheels are cast magnesium with knock-off hubs.

A de Dion rear end has at last reappeared in this country. Some R&T readers may not know the principle or its advantages; the others will please excuse us while we give the reasons why Chevrolet has taken this route. Briefly (and oversimplifiedly), the differential is rigidly attached to the frame, which is itself sprung on a tubular member that connects the wheels. The unsprung weight, which is low, can be controlled better than with other systems, and engine torque is absorbed within the frame, not affecting rear wheel loading. The differential moves with the frame, so there are 4 universals.

Horsepower	300 plus
Rear axle	de Dion, quick change differential
Transmission ratios (all forward gears synchronized)	
4th	direct
3rd	1.22
2nd	1.54
1st	1.87
Body	Magnesium, hinged at cowl
Frame	Chrome molybdenum steel tubing, weight 180
Front suspension	Independent coil spring, variable spring rate, link-type stabilizer
Rear suspension	de Dion bar with 4 articulating arms, coil springs, variable spring rate. ●

ROAD TEST 4-SPEED CORVETTE

IN JANUARY of this year, Chevrolet announced that their Corvette would be available not only with fuel injection, but also with four speeds forward. We were shown the transmission unit at the time (in Detroit) but only recently have they become generally available.

Prior to testing the 4-speed model we were given a standard 3-speed fuel-injection car to drive, and though a complete road test was not made, we did publish some of the data last month. Unfortunately, exact comparison of data between the two cars is not possible because the 3-speed car had a 3.70 axle, standard differential, stock rear springs and original equipment (6.70-15) tires.

Our 4-speed test car is the property of enthusiastic owner-driver Andy Porterfield (his 3rd Corvette). The car was raced at Palm Springs and Santa Barbara a month earlier and had just had the new gearbox installed on the day before the test. A standard Chevrolet option, "Positraction" rear end with 4.11:1 gears, was installed, along with heavy-duty rear springs (one extra leaf), stiffer shocks and special re-treaded 6.50-15 tires. We measured these tires very carefully; as loaded and inflated they are a close equivalent to the 7.10-15 size, which give 720 tire revolutions per mile at 35 mph and 2960 engine revs per mile. The 3-speed car, with radio and heater, the hard top and full tank weighed 2880 lb, while the 4-speed car without radio and heater and with cloth top weighed exactly the same.

Even the Anglophiles now readily admit that the Corvette will go. The only question left is how *well* it goes. Our figures in the data panel are, as usual, the mean of several runs in opposite directions, and corrected for speedometer error. Owner Porterfield, who drove during the tests, made consistent starts with very little trouble and brisk shifts without speed-shifting. The data are unequalled by any other production sports car.

In 1st gear, the normal production 3-speed car was an abomination, for the standard rear springs are much too flexible and the standard tires without benefit of the special differential make wheelspin difficult to overcome. Under the same conditions the 4-speed car, with its modifications, was nearly fault-free, although initial wheelspin could be provoked. When this was done, the rear end would chatter and hop for the first 50 ft.

Either car would reach 60 mph from a standstill in one magnificent roar, using 1st gear only. In the case of the 4-speed car, the 4.11 axle meant that 6600 rpm was required, but the engine seemed quite willing to go on beyond that. The owner states that he exceeded 7000 rpm on several occasions when racing with the original 3-speed gearbox (which has the same gear ratio in 1st as the new box).

A study of the acceleration chart would indicate that the 2nd-gear spurt is too short (from 60 to 76 mph) but the rev limits used for 1st and 2nd were not the same (see data under

add fuel injection and get out of the way

performance). A truer picture of useful shift points can be gathered by considering the fact that even 1st gear is synchronized. In a race, it would normally be used on a slow turn from, say, 41 mph to 56 mph, while if 2nd is used over an equal rpm range (from 4500 to 6200 rpm) the corresponding speeds are 56 to 76 mph. In other words, the gear ratios are close and very well spaced for competition work. Incidentally, although the gearbox was brand new it shifted beautifully, with an easy, short throw. It is virtually impossible to clash and is also much quieter than the 3-speed box. Though we approached the synchronized low with caution it didn't complain.

The fuel-injection engine is an absolute jewel, quiet and remarkably docile when driven gently around town, yet instantly transformable into a roaring brute when pushed hard. It idles at about 900 rpm and pulls easily and smoothly from this speed even in high gear. Its best feature is its instantaneous throttle response, completely free of any stutter or stumble under any situation. The throttle linkage has a certain amount of back-lash and friction, but there are no flat spots such as we described in last year's test of two Corvettes with twin 4-barrel carburetors.

No timed top speed runs were made for the simple reason that it would be relatively pointless to do so, with the short-course 4.11 axle gears. An engine speed of 6500 is easily reached in 4th gear, equivalent to 132 mph with no allowance for tire expansion. With suitable gears the Corvette can approach 150 mph, as has been proven at Bonneville and at Daytona.

The brakes on our test car were normal duo-servo Bendix type, but with Ferodo linings. The owner reports good though not ideal anti-fade characteristics; we made three stops from just over 100 actual mph with no sign of trouble and a normal feel. A brake booster is not supplied, nor is it needed.

Perhaps a few words should be allotted to the numerous items of special equipment which are available. The base price of the Corvette is $3909.52 f.o.b. St. Louis. This includes the fuel-injection engine with high-performance camshaft, lightweight valves, etc., but very little else. The 4-speed transmission is $188 extra, or $275 outright. Wheels with extra wide rims (5.50 in.) are $14.00 each.

In addition, there is a special package bearing part No. RPO-684, which carries a list price of $725. Included are:

1. Heavy-duty front springs which increase the ride rate from 105 to 119 lb/in. (14% stiffer).

2. Special heavy-duty rear springs with a ride rate of 125 lb/in., not 115 (8.7% stiffer). (continued on page 29)

CHEVROLET CORVETTE
with four-speed gearbox and fuel injection

SPECIFICATIONS

List price	$4098
Wheelbase, in.	102
Tread, f/r	57.0/59.0
Tire size	6.50-15
Curb weight, lb	2880
distribution, %	53.5/46.5
Test weight	3180
Engine	V-8, ohv
Bore & stroke	3.875 x 3.0
Displacement, cu in.	283
cu cm.	4639
Compression ratio	10.5
Horsepower	283
peaking speed	6200
equivalent mph	134
Torque, lb-ft	290
peaking speed	4400
equivalent mph	95
Gear ratios, overall	
4th	4.11
3rd	5.38
2nd	6.82
1st	9.28

CALCULATED DATA

Lb/hp (test wt)	11.23
Cu ft/ton mile	152.4
Engine revs/mile	2960
Piston travel, ft/mile	1480
Mph @ 2500 ft/min	101

PERFORMANCE, Mph

Top speed (at 6500 rpm)	132
3rd (6500)	101
2nd (6500)	80
1st (6600)	60
Mileage range	11/16 mpg

ACCELERATION, Sec.

0-50 mph	4.7
0-60 mph	5.7
0-70 mph	7.7
0-80 mph	10.2
0-90 mph	13.3
0-100 mph	16.8
Standing start ¼ mile	14.3

TAPLEY DATA, Lb/ton

4th	380 @ 82 mph
3rd	505 @ 69 mph
2nd	off-scale
1st	off-scale
Total drag at 60 mph, 119 lb	

SPEEDOMETER ERROR

Indicated	Actual
30 mph	28.8
40 mph	37.4
50 mph	47.1
60 mph	57.5
70 mph	67.0
80 mph	77.4
90 mph	87.8
100 mph	98.2

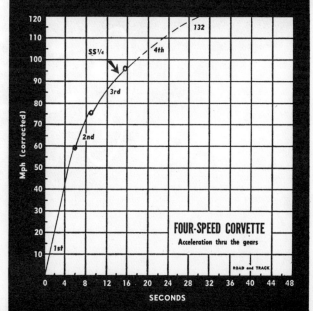

FOUR-SPEED CORVETTE
Acceleration thru the gears

ROAD & TRACK
THE MOTOR ENTHUSIASTS' MAGAZINE

1958

December 1957

In the swim with dual headlights, the new model has grown too fussy.

1958 CORVETTE

Sportsmen and comfort-lovers can fit it to either personality, or both

That supposedly hard-to-sell commodity, elegant simplicity, is gone.

A beautiful wheel, crying for instruments and switches to match . . .

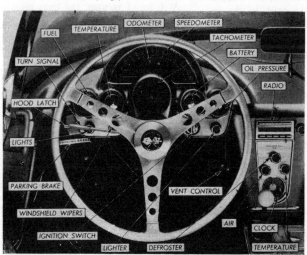

THEY SAY that the functions of the automotive press social season, now in full swing, fall into two major categories: those where the liquor is noteworthy but the car is not, and vice versa. Unfortunately, the same does not apply to the first flow of gilded rhetoric from the advertising agencies in regard to the latest products.

This month's cover car has been the subject of sundry improvements, as well as the corrosive influence of the "stylists." It can stand firmly on its still-15-inch wheels without the assistance of copywriters' ecstatic groans, from which we have gleaned the following facts:

Chevrolet has gone farther than ever before in attempting to please, via the option route, all potential Corvette buyers. Those who want a real sports car will lean toward the dual 4-barrel carburetors or the fuel-injected engine, with or without the special camshaft; compression ratios go up to 10.5:1, and an 8000-rpm tachometer comes with the top performance combination. A heavy-duty clutch and a non-slip differential (with either the standard close-ratio 3-speed gearbox or the all-synchronized 4-speed transmission; the latter was tested by Road & Track in the August issue) will also please the driving-minded buyer. So will optional heavy-duty suspension, with spring rates of 340 at front and 125 at rear instead of the standard 300 and 115; 1⅜-inch-diameter shocks instead of 1 inch; a 1³⁄₁₆-inch roll bar instead of 1¹⁄₁₆; and an adapter for heavy-duty steering (also optional), whose 16.3:1 overall ratio replaces 21.0:1.

The cockpit, whose features are identified for the feeble-minded in the photo at left, is a highly satisfactory job of pleasing buyers with different tastes. The most glaring fault of previous models, the too-small tachometer placed low in the center, has been rectified. It is now in front of the driver, where it belongs, though it is still too small.

The first mass-production use of a central "control tower" is here haltingly pioneered. No true car controls are mounted in this spot, which has the advantage of their not being reachable, on purpose or accidentally, by a passenger. There are true gauges for oil pressure and generator charge. A broad, vinyl-covered grab bar takes up most of the right-hand side of the dash. Reflectors on the inside of·the door warn oncoming drivers at night. A transistor radio and a new hard top (see cover) are extras, as is a power-operated soft top.

TUNE-UP CLINIC

by BILL COREY

CORVETTE FUEL INJECTION

ON PAPER, the Corvette injection system seems to be nothing more than a "dribble" carburetor with jets located at each intake port instead of in a carburetor body. Experience has proved how wrong this judgment is. The system works to near-perfection and eliminates many of the bothers with one or two four-throat carburetors. In 1957, the factory made certain changes which were carried over to the '58 series: the addition of a spark port to adapt the injection to a vacuum advance distributor, and a revised cold enrichment assembly. These changes make late cars idle quite well, and give better fuel economy, better hot and cold starting and improved part-throttle acceleration.

Servicing of the injection system requires no magic wand or special tools, mainly because field service must be confined to adjustments to the idle fuel and air circuits, cold enrichment rod length, and cold enrichment coil index setting. Under no circumstances may the main control diaphragm cover, fuel meter bowl cover, or ratio stops be removed or tampered with. If they are, the only solution is to replace the fuel meter–injector nozzle assembly with new parts.

As with any injection system, cleanliness of fuel *and* air supplies is of extreme importance. The first step in tuning is to clean the air filter element thoroughly. Replace it if it has been used over 15,000 miles, or if you doubt its ability to pass air freely. The fuel filter is next; it should be replaced semi-annually. To remove the element, take off the filter cover and insert three pieces of 0.040-inch shim stock between the element and the clips inside the filter. The element can then be removed by pulling up. After filters have been cleaned and electrical tuning completed, start the engine and allow it to warm up until the throttle tab is completely off the fast-idle cam. At this point, you are ready to start adjustments.

If a new fuel injection unit has been installed or if the unit has been serviced in any way, it is best to start by fully closing both the idle and idle fuel adjustment screws. Both screws are then turned out two turns. At this point, readjust the air screw for proper idle speed; then readjust the fuel screw for smoothest operation. When this is done, there will no doubt be a change in engine speed. So, after you have the best idle, readjust the air to obtain a 500–revolutions per minute setting. Repeat your fuel adjustments for smoothest idle. It is permissible to work back and forth with these adjustments to achieve the smoothest running at the specified 500 rpm.

Choke or cold enrichment adjustments are checked by setting the scribed mark on the choke thermostat housing 1½ notches rich. The cold enrichment rod adjustment must be made with the engine off and cool. Remove the rubber sleeve from the housing signal boost tube and install a short length of wiper hose over the tube. Then open the throttle just enough to allow the throttle tab to rest on the fast-idle cam, and hold the trip lever manually against the counterweight tab. Blow gently through the end of the wiper hose. If a slight air flow can be heard at the air meter, the choke rod length is correct. Repeat this check with the throttle tab on the second step of the fast-idle cam. Here, no air flow should be heard. To adjust, shorten or lengthen the rod by bending. Then start the engine and warm it thoroughly. Place the throttle tab manually on the fast-idle step of the cam. Engine speed should be 1600 to 1700 rpm. If not, bend the throttle tab in or out to correct it.

TROUBLE SHOOTING

In order of importance, here are the causes of most malfunctions of the system: 1) Dirt: fuel and air supplies must be kept thoroughly clean; inspect parts thoroughly. 2) Wear: watch for wear in vital moving parts; if in doubt, replace. 3) Maladjustment: check all adjustments carefully. 4) Air leaks: examine all gaskets, tighten all manifolds, check all casting surfaces. 5) Distortion: watch for parts that are overtightened or bent out of shape.

In practice, we find that air leaks are most important to guard against, just as with supercharging. To check for leaks in the main control diaphragm, disconnect the vacuum signal line at the end opposite the diaphragm connection and attach a hose from a vacuum source and vacuum gauge to the tube. (Most distributor test machines have such a source and gauge.) With the gauge's vacuum release valve shut, apply and hold a vacuum on the diaphragm and watch the gauge dial. If there is a leak, the needle will slowly slip to a lower reading, and the fuel meter must be replaced. (When testing, disconnect the vacuum signal line from the opposite end of the T and install a plug.) Incidentally, never apply a vacuum of more than 4 in. to the diaphragm of the main control diaphragm. The enrichment diaphragm, on the other hand, should be checked with 12-16 in. of vacuum. These specifications must not be exceeded even momentarily. It is, of course, important that no leaks exist in the test rig before replacing parts. ⊙

CORVETTE FUEL INJECTION
MORE ON TROUBLE SHOOTING

WE OFTEN have complaints about hard starting of fuel injection engines. If this is your trouble, first be sure that you are following correct starting procedure. If the engine is cold, the accelerator should be depressed just once to set the choke fast-idle cam. Then release the pedal completely until the engine starts. If there is still difficulty, make sure that the hot-starting micro switch is properly adjusted. This should not be actuated by the throttle cam until ¾ throttle is reached. Bend it to adjust if necessary. The next item to check is the starting solenoid on the fuel meter. This should operate when the starter motor is engaged with a closed throttle. If it is inoperative, check the starting cut-off switch and solenoid. Next, see that fuel is flowing to the fuel distributor by loosening the line at the fuel meter while the engine is being cranked. If fuel does not flow, the fuel valve is sticking and must be cleaned. Push the solenoid plunger to free the fuel valve, and clean it thoroughly. Another possibility is a clogged fuel line to the distributor. To check this, remove one set of nozzles from a nozzle block and check the flow while cranking the engine. If there is no flow, suspect the fuel distributor check valve or a clogged fuel-meter-to-distributor fuel line. If fuel is present at all places and the electrical system is all right, check for a massive air leak, which may be caused by loose or cracked nozzle blocks as well as the more conventional reasons. In extreme cases, you may have to disconnect the enrichment line at the choke housing to provide full enrichment. If this eliminates the trouble, the valve in the housing is not seating properly. Clean it, or replace the defective housing. Insufficient fuel pressure is another possible cause of failure to start when cold. This should be 4¾ to 5½ pounds per square inch.

Flat spots in fuel-injection engines are unlikely, but if they exist check for vacuum leaks in the signal lines and fittings, and check the main control diaphragm venturi signal passage and auxiliary signal passages in the air meter for cleanliness. The main control diaphragm T should be clear, and there should be no sticking in the spill plunger. Diaphragms should be checked with a vacuum gauge as described before. Finally, check the enrichment control diaphragm rod length to be sure that it allows proper cut-in for power and economy. Adjust it by connecting a vacuum source and gauge to the enrichment vacuum line. Apply and hold a vacuum of 12 to 15 in.; then slowly release it, noting the gauge readings as the lever leaves the economy stop (forward) and arrives at the power stop (rear). If this adjustment is correct, the rod length will allow the lever to reach the power stop at 3 in. of vacuum (or above) while leaving the economy stop at 9 in. (or below). At 6 in. of vacuum, the lever must not be touching either stop. Adjust the rod length by removing the enrichment diaphragm cover. Of course, you must have established that the enrichment diaphragm is in functioning order and receiving engine vacuum from the cold enrichment housing. If no vacuum is present, look for trouble in the housing such as broken heat element posts, a burned-out heat element, or a stuck ball in the enrichment valve.

As a general rule, fuel-injected engines give somewhat better fuel economy than their carbureted brothers. However, economy depends upon proper functioning of all of the signal devices in the system. If gas mileage is down, first suspect improper adjustment of the enrichment lever and see that it rests on the economy stop after the engine has been in use or warmed up for five minutes or more. When it is completely off fast idle, also check that there is no signal (vacuum) at the signal boost tube. Do this by removing the rubber sleeve and putting your finger over the tube. If you feel suction, the signal boost valve in the cold enrichment housing is leaking and should be cleaned. It is also important that *accurate* manifold vacuum signals are reaching the enrichment diaphragm. First check the actual engine manifold vacuum; then make the same check at the signal end of the vacuum line at the enrichment housing. Readings should be within 1 in. If they are not, suspect leaking gaskets, a stuck valve, or a warped air meter fit at the enrichment housing. As a final check on economy settings, inspect the ratio stop screw adjustments. These stops are pre-set at the factory and should never be altered in the field. The lock nuts are coated with a blue sealer. If it has been disturbed, the fuel meter assembly must be replaced. The ratio stops can be set only at the Chevrolet factory.

CHANGES FOR 1959

Specifications of 1959 Corvette engines vary only in detail from those used in 1958 and earlier V-8 cars.

One displacement size, 283 cubic inches, is standard on all models; bore is 3⅞ in., stroke 3 in. Firing order is 1-8-4-3-6-5-7-2. No other engine sizes are optional with the Corvette, as the 348 engine has not yet proved to be as versatile as the 283. Five 283 engine options are available, however, to suit all tastes from show to go. The smallest horsepower rating is 230. This engine uses a single four-throat carburetor, dual exhausts are stock, compression ratio is 9.5:1 and hydraulic tappets are combined with a camshaft having .3987-in. lift for both intake and exhaust. Fuel of 94 to 96 octane rating is specified and static ignition timing is 4° before top dead center at an idle speed of 475 revolutions per minute. If the car has the Powerglide transmission, set the timing at the same number of degrees, but reduce the speed to 450 rpm. The cam used is timed to open intake valves at 12° 30′ BTDC and close them at 57° 30′ ABDC. This is mild by present-day standards, and the exhaust timing is not radical either, at 54° 30′ opening and 15° 30′ closing. (Very similar to Jaguar valve timing, by the way.) Spark plug gap is .033 to .038 in.

The next engine option is the 469-A (or B), which is identical in all respects except for the dual four-throat carburetors. This brings rated bhp to 245. However, one small change in the 469-C engine steps up bhp to a whopping 270. That change is the camshaft. Timing is as follows: intake valve opens at 35° BTDC, closes at 72° ABDC; exhaust valve opens at 76° BBDC, closes at 31° ATDC. With this cam, clearances are specified at .012 in. for intake and .018 in. for exhaust (hot).

Fuel-injection engines are the 579-A or C and the 579-D. The former have specifications identical to the 469-A and B and are rated at 250 bhp. The 579-D is the twin of the 469-C except for a 10.5:1 compression ratio and the fuel injection. This adds up to 290 bhp, and 98 to 100 octane fuel is a must. No change in static timing is required on the dual four-throat 469-A and B engines or with the fuel-injection type with the milder cam and hydraulic lifters. However, when the hot cam is used with twin carburetors, timing should be set at 7° BTDC at 800 to 850 rpm; if fuel injection is combined with the cam, 14° BTDC at 650 rpm is specified.

The distributor on all models calls for a point gap of .018 in. when new points are installed. If old ones are being reset, Chevrolet suggests .014 in. Breaker spring tension is 19-23 ounces. ◉

CORVETTE TEST
CONTINUED FROM PAGE 25

3. Special shock absorbers with a working diameter of 1.375 in. instead of 1.0.

4. Special stabilizer bar with diameter increased from .6875 to .8125 in.

5. Fast steering adapter, a plate which changes the overall steering ratio from 21.0:1 to 16.3:1 (Here Chevrolet seems confused, since they advertise 16.0:1 as standard steering ratio. However, this adapter reduces turns from 3.7 to 2.9)

6. A Spicer "Positraction" differential and choice of 3.70, 4.11 or 4.56 gears.

7. Finned cast-iron brake drums and vented backing plates with air scoops.

8. "Cerametalix" brake linings.

Chevrolet said, back in 1954, that they were in the sports car business to stay, and their competition successes of the past two years certainly show that they meant it. ●

ROAD & TRACK

MISCELLANEOUS RAMBLINGS

BY JOHN R. BOND

New Corvette

While in Detroit recently, I had a long talk with Zora Arkus-Duntov, the man most interested in Chevrolet's Corvette. He's quite enthusiastic about the aluminum cylinder heads, a new option for 1960. Compression ratio has been raised to 11.0:1 and this, in conjunction with other changes to improve breathing, boosts the fuel injection version from 290 to 315 bhp. Intake valves are larger— 1.94 in. instead of 1.72—and the injection manifold now has a larger plenum chamber over the ram pipes. The heads save 53 lb of weight, bringing the engine total down to about 480 lb without flywheel or clutch. Incidentally these heads are cast of a new material (presumably one of the new extra-high-silicon alloys) and do not use valve seat inserts. Material for the clutch housing has been changed to aluminum also, saving 18 lb. Interesting options will be a new all-aluminum cross- flow radiator and a temperature-modulated, variable-speed fan drive, which limits fan speed to 3100 rpm.

Except for a few minor interior changes, the Corvette's appearance is the same as for 1959. But underneath the car there are some very significant suspension modifications. The more obvious changes are a heavier front anti-roll bar and the addition of a similar bar at the rear, just ahead of the axle. The most significant change is an apparently innocent one: an increase of one in. in rear-spring rebound travel. Zora assures us that this new set-up is extremely effective and that the new car with standard "soft" suspension will definitely out-handle and out-corner the 1959 heavy-duty option. In fact it's so good that they have discontinued RPO-684, the heavy-duty springs and dampers. There is also a modification in pinion nose angle to give a smoother- operating drive line.

As before, there are two brake options, RPO-686 (sintered-iron linings) and RPO-687 (the ceramic-metallic type). Larger front-wheel cylinders put more power into the front wheels with slightly reduced pedal effort.

The high-performance engines (intended primarily for racing) are given very specialized treatment. In addition to customary inspection, many critical parts are now routed through a special department for a very painstaking examination of dimensions, flaws, finish and quality of materials. Included in this group are valves, rocker arms, pushrods, pistons, connecting rods and crankshafts. Just a few years ago Chevrolet would have laughed at such a suggestion. This certainly shows how serious they are about the sports-car side of the business.

ROAD TEST 1959 CORVETTE

A pretty package with all the speed you need, and then some

CHEVROLET is now entering the sixth year of production on its famous and highly respected Corvette. Since its introduction as the 1954 model, the car has changed so much that it is hard to think of it as the same make.

Body configuration and general styling were the same for the 1954-55 models and for the 1956 and 57's. Now the 1959 version shares the body used on the 1958 Corvette.

The 1954 model was rather prosaic and had little to offer with the exception of being different from the other U.S.-built cars. Powered by a 6-cylinder engine that was essentially the same as the one introduced in the 1937 Chevrolet, the Corvette offered neither performance nor exceptional appearance. In 1955 the fiberglass-bodied car started to come alive when Chevrolet's new V-8 engine was offered as an option.

In 1957 the performance potential was given a real boost when fuel injection and a 4-speed, all-synchromesh transmission were made available. A heavy-duty suspension kit (developed at the request of competition-minded enthusiasts) included stiffer springs, dampers and improved brakes. These were all retained for 1958, and the commercial artists inflicted quad headlights and fake hood louvers on the only production sports car built in the U.S.

The new car will not necessarily make owners of the 1958 Corvette rush out to buy a new one, but a few worthwhile changes have been made.

Looks, while not the most important factor to consider, are certainly the first thing noticed about any car. The appearance of the 1959 Corvette has been improved by the simple expedient of removing the phony hood louvers and the two useless chrome bars from the deck lid.

Inside, the most significant change, and one of the only two that are noticeable, is in the seats. These have been redesigned and are among the most comfortable seats in any car, sports or otherwise. They quite adequately do the job of holding driver and passenger comfortably in place during all but the most violent action. We do feel, though, that safety belts would be desirable if much hard driving is to be done.

The second obvious change in the interior is the addition of a fiberglass package tray under the passenger's "grab rail." This tray serves in the usual glove compartment capacity as a catch-all for odd bits of useless detritus and

Cleaner hood lines (from simply eliminating the fake hood louvers) have made a slight improvement in looks for the 1959 Corvette. Trim on Corvettes, like all GM cars, is extremely well executed whether it is functional or mere decoration.

Even if the Corvette does carry more embellishment than most sports cars, it is still a sleek-looking package.

in this case (if the top is down), leaves, dust and possibly rain water.

This package tray is the first item that should be removed from a Corvette, as this badly placed receptacle also presents a safety hazard to the passenger's knees. The second item to be removed would be the grab rail itself: the thinly padded bar would probably do more harm than good in most cases.

We were fortunate, through the efforts of Steve Mason, to be able to use the facilities of Riverside Raceway to conduct the Corvette test. Frank Milne of Harry Mann Chevrolet (who furnished the test car) accompanied our test crew to the site and assisted in assembling the data on page 25.

The first item on the check list, and one on which all results would hinge, was speedometer error. The mile-long straight at Riverside has a measured quarter and half mile which made this check simple and quick to perform. After taking the readings in 10-mile increments from 30 to 100 miles per hour we discovered the error to be about average for most cars tested, with the maximum error of 4 mph indicated at 100 mph.

Tapley readings, as can be seen in the data panel, were somewhat lower than those obtained with a similarly equipped 1957 Corvette (Road & Track, August 1957). The subsequent acceleration tests indicated the newer car to be somewhat slower getting off the starting line, but the figures improved as the upper speed ranges were reached. This could be explained by the additional weight of the 1959 model and by the newness of the car (it had less than 500 miles on the odometer) when we took it out for the test.

Starts were brisk but made with very little wheelspin, due to the efforts of the Positraction rear end, causing the engine to bog down just off the line. The first 20 yards were not very impressive as the engine struggled to overcome the lack of low-speed torque.

This is not to imply that the Corvette does not have torque. It does, but with a combination of peak torque at 4400 revolutions per minute and peak horsepower at 6200 rpm, it does have a lag before the engine really gets to work. This initial hesitation can be seen on the acceleration graph.

The 4-speed, all-synchromesh gearbox is still one of the smoothest-working units it has been our pleasure to use. (It is now optional on all Chevrolet V-8's.) The lever is placed at just the right location for driver convenience, and shifts, either up or down, can be made quickly and easily at any speed within engine limits. The main factor to consider in downshifting is the possibility of over-revving the engine in the process.

A new feature this year is a positive lock-out for reverse gear. This simple pull handle on the gearshift lever can be actuated by the first two fingers of the right hand as it rests on the shift knob.

Weight distribution of the car had shown 53% of the weight to be on the front wheels with one person in the car, but in spite of the nose-heavy attitude there was a marked tendency to oversteer. It was found with a little practice that a drift, once set up, could be maintained with little effort by dextrous manipulation of the throttle and steering wheel. The abundance of horsepower at the driver's command and adequately quick steering helped somewhat, too.

An extremely comfortable, though not very capacious, interior has new seat contours.

Heart of the Corvette is the fuel-injected, 290-horsepower, 283—cubic inch V-8 engine.

PHOTOGRAPHY: BATCHELOR

Panel lifts with top to disclose more luggage space.

The test car was equipped with the optional suspension (stiffer springs and dampers) which contributed to the ability to negotiate curves at maximum speed. Also installed on the car as part of the suspension kit was the racing brake set-up which includes Cerametalix lining, finned drums and air ducts for additional cooling. The brakes proved completely satisfactory during the test, with no fade being evident, and brought the car down from high speed in a straight line on every application.

The Cerametalix lining is not as effective when cold as it is when hot, and after the test was completed the lining squeaked when the brakes were applied. In spite of these minor annoyances, the brake kit is to be recommended for anyone contemplating fast driving with a Corvette and is an absolute necessity for competition work.

Driving this course at high speeds also confirmed our earlier opinion of one definite advantage of fuel injection. It is claimed by many Corvette enthusiasts that a Corvette equipped with dual 4-barrel carburetors will outperform the FI model at the upper end. This may or may not be true, but the fuel injection has it all over carburetors for throttle response and lack of sensitivity to motion. There is no flooding or starving on hard cornering with fuel injection.

Taking everything into consideration, the Corvette is a pretty good car. It probably has more performance per dollar than anything you could buy and parts are obtainable without sending to Italy, Germany or England.

The changes to the car in the last six model years are not so great as we think will come about in 1960. We predict that this will be the year of the big changes for Corvette, and most of them for the better.

Trunk space is larger than that of most sports cars but still requires owner to travel light.

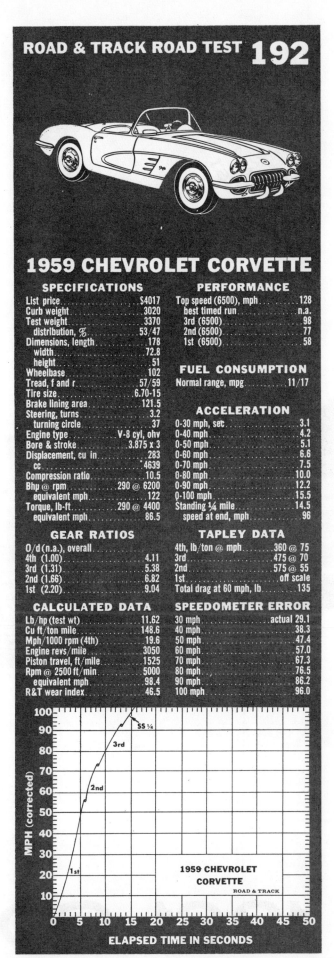

ROAD & TRACK ROAD TEST 192

1959 CHEVROLET CORVETTE

SPECIFICATIONS

List price	$4017
Curb weight	3020
Test weight	3370
distribution, %	53/47
Dimensions, length	178
width	72.8
height	51
Wheelbase	102
Tread, f and r	57/59
Tire size	6.70-15
Brake lining area	121.5
Steering, turns	3.2
turning circle	37
Engine type	V-8 cyl, ohv
Bore & stroke	3.875 x 3
Displacement, cu in	283
cc	4639
Compression ratio	10.5
Bhp @ rpm	290 @ 6200
equivalent mph	122
Torque, lb-ft	290 @ 4400
equivalent mph	86.5

GEAR RATIOS

O/d (n.a.), overall	
4th (1.00)	4.11
3rd (1.31)	5.38
2nd (1.66)	6.82
1st (2.20)	9.04

CALCULATED DATA

Lb/hp (test wt)	11.62
Cu ft/ton mile	148.6
Mph/1000 rpm (4th)	19.6
Engine revs/mile	3050
Piston travel, ft/mile	1525
Rpm @ 2500 ft/min	5000
equivalent mph	98.4
R&T wear index	46.5

PERFORMANCE

Top speed (6500), mph	128
best timed run	n.a.
3rd (6500)	98
2nd (6500)	77
1st (6500)	58

FUEL CONSUMPTION

Normal range, mpg	11/17

ACCELERATION

0-30 mph, sec	3.1
0-40 mph	4.2
0-50 mph	5.1
0-60 mph	6.6
0-70 mph	7.5
0-80 mph	10.0
0-90 mph	12.2
0-100 mph	15.5
Standing ¼ mile	14.5
speed at end, mph	96

TAPLEY DATA

4th, lb/ton @ mph	360 @ 75
3rd	475 @ 70
2nd	575 @ 55
1st	off scale
Total drag at 60 mph, lb	135

SPEEDOMETER ERROR

30 mph	actual 29.1
40 mph	38.3
50 mph	47.4
60 mph	57.0
70 mph	67.3
80 mph	76.5
90 mph	86.2
100 mph	96.0

Graph: MPH (corrected) vs ELAPSED TIME IN SECONDS — 1959 CHEVROLET CORVETTE, ROAD & TRACK. Curve marked 1st, 2nd, 3rd, SS ¼.

America's competition-proven sports car revisited

1961 CORVETTE

ONCE UPON A TIME, just a few years ago, owners of America's only sports car were on the receiving end of constant gibes from the "sporty car set," which held that the only thing the beast had to offer was drag strip performance. It would go like the wind (in a straight line, they said), but it wouldn't corner, it wouldn't stop, it had a boulevard ride, and a glass body. And it took 265 cu in. (4.5 liters) to get that performance.

Well, these derogatory remarks probably were true at one time. At least, some of them were. But Chevrolet engineers have now achieved an excellent package, combining acceleration, stopping power, a good ride and handling characteristics whose adequacy is indicated by the car's race-winning ways.

In our January 1959 test report of the 1959 Corvette we said that 1960 would be the year for big changes in the Corvette. We were wrong. The 1960 model wasn't too much different from, or too much better than, the 1959 version. Lacking any great changes in 1960, we might logically have predicted a major change in 1961, but luckily we didn't.

However, the few changes which have been made are for the better. Continual refinements since 1954 have made the Corvette into a sports car for which no owner need make excuses. It goes, it stops, and it corners.

The major change in the appearance is the rear end treatment, which was derived from the Sting Ray, GM racing Corvette, owned by Bill Mitchell. The stubbier look achieves a more crisp and a fleeter appearance than that of previous models, which looked "soft." The front end remains basically unchanged. New bumpers fore and aft blend nicely into the body design, and the exhaust tips are now under the body instead of through the bumper tips. This was a good move; there's no mistaking the Corvette for any other make and it is a better looking car now.

The finish of the fiberglass body is generally excellent, although we did find a few minor flaws on our test car, mostly in obscure places. Panel fit and fairing from one panel to another were good and showed Chevrolet's great attention to the Corvette molds.

Interior trim and design are similar to past models and well done, but have a little of the Motorama touch. The seats are excellently designed and are very comfortable. Our longest single excursion was of some 200 miles, but no sign of driver fatigue was evident and we honestly feel a day behind the wheel of a Corvette could be put in without undue strain.

The instruments are easy to read and include a speedometer, tach (reading to 7000 rpm—red-lined at 6200), gas gauge, temperature gauge, oil pressure gauge and ammeter. Indicator lights are used for the turn signals, high headlight beam and parking brake. The parking brake light on the panel lights up when the key is turned on (if, of course, the parking brake is on) and when the engine is started the light blinks its warning to the driver. The only fixture in the Corvette interior that's hard to use is the radio, which is mounted in a console deep under the instrument panel. This console also carries the clock, which is difficult to see and should be looked at only when the car is stopped.

The seats have 3 in. of fore and aft movement, which gave everyone on our staff adequate leg room, but the body panel between the seats interferes with the driver's elbow (when shifting) when the seat is at its rearmost position.

Vision in the Corvette is excellent when the car is equipped with the hardtop. Slim pillars and lots of glass area are responsible. We used two test cars, one white and one metallic blue. Reflections from the white rear deck into the rear window of the white car caused a hazy view when we looked out of the back window or used the rear view mirror. The darker colored car did not give us this trouble. Even in the white Corvette no haze was noticeable in the windshield, because the cowl was covered with red material to match the upholstery.

We noticed a considerable amount of wind noise with the windows rolled down, and if the windows are rolled up they have to be all the way up or the driver feels a bad draft. Some engine and wind noise was evident even with both windows up.

The car started with a flick of the key on cool morn-

ings but, oddly, it was difficult to start on several occasions after the car had been thoroughly warmed up. This is unusual and could most likely be cured by further tuning.

The clutch operation proved extremely smooth, whether we were plugging along in slow-moving traffic or getting off the line for a standing start acceleration run. We tried a stop and start on a hill of approximately 30% grade and found that by dextrous manipulation of the clutch and throttle it could be surmounted with ease and smoothness. We hadn't been worried about a lack of horsepower, but we had wondered about the flexibility of the engine until we tried the hill.

Once more we found the injector to be extremely flexible. The engine is able to pull smoothly from under 15 mph to top speed with no bucking or hesitation. In our test car, which had 4.11:1 rear axle gears, it was much easier, of course, than it would have been with the top ratio available. Flexibility, rather than actual increased power, is the main justification for the injectors and in road racing use they allow full power to be maintained under all conditions. Carburetors will almost invariably starve when the car turns one way and flood

when it corners in the opposite direction; and quite often they will flood under extreme deceleration, making it difficult for the driver to accelerate properly out of the turn.

Options for 1961 include five horsepower ratings: 230 (standard), with a single 4-barrel carburetor; 245, with twin 4-barrels; 270, with twin 4-barrels; 275, with fuel injection; and 315, with fuel injection. Three transmission options are available: 3-speed synchromesh; 4-speed synchromesh, and Powerglide. Five rear axle ratios are available: 4.56:1, 4.11:1, 3.70:1, 3.55:1 and 3.36:1.

Obviously, any buyer should find a combination to suit his needs. Performance of this Corvette was little different from the one we tested two years ago and only improved at the upper end. A similar car, owned by Alan Lockwood, race tuned, timed 107 mph at the end of a quarter mile at the LADS drag strip, Long Beach, Calif.

We've said many times that we think the Corvette 4-speed transmission is one of the best in the world and we have no reason to change our minds. In all, five different people on our staff drove the car and none was able to fault the synchromesh. The ratios are excellently spaced and, with synchromesh on all four gears, the driver always has the proper ratio at his disposal by a mere movement of the lever.

We were greatly impressed by the combination of a very good ride coupled with little roll on corners. Most cars with riding qualities approaching those of the Corvette can't match its sticking ability on curves. And those that match or beat its handling usually ride like the proverbial truck.

The Corvette is absolutely unmatched for performance per dollar in terms of transportation machinery (some of the newer Formula Junior cars will beat it for performance per dollar, but are, of course, single purpose cars).

The steering is accurate, though a little slower than we would like in a car with this much power and speed. The brakes proved up to every test we put them to and a sudden panic stop to avoid a day-dreaming motorist increased our admiration for the refinements and improvements made by Chevrolet engineers in the Corvette since its introduction. Once again we want to thank Harry Mann Chevrolet for furnishing the test car. The following weekend the Corvette was used as the pace car at the Riverside Sports Car GP.

ROAD TEST NO. 270
1961 CHEVROLET CORVETTE

SCALE: 10" DIVISIONS

DIMENSIONS

Wheelbase, in	102
Tread, f and r	57/59
Over-all length, in	178
width	70.4
height	52.1
equivalent vol, cu ft	378
Frontal area, sq ft	20.4
Ground clearance, in	6.7
Steering ratio, o/a	21.1
turns, lock to lock	3.7
turning circle, ft	37
Hip room, front	49
Hip room, rear	
Pedal to seat back	38
Floor to ground	14

CALCULATED DATA

Lb/hp (test wt)	10.7
Cu ft/ton mile	148.3
Mph/1000 rpm (4th)	19.6
Engine revs/mile	3050
Piston travel, ft/mile	1525
Rpm @ 2500 ft/min	5000
equivalent mph	98.4
R&T wear index	46.5

SPECIFICATIONS

List price	$3872
Curb weight, lb	3080
Test weight	3390
distribution, %	53/47
Tire size	6.70–15
Brake lining area	157
Engine type	V-8 cyl, ohv
Bore & stroke	3.88 x 3.0
Displacement, cc	4639
cu in	283
Compression ratio	11.0
Bhp @ rpm	315 @ 6200
equivalent mph	122
Torque, lb-ft	295 @ 4700
equivalent mph	96.1

GEAR RATIOS

4th (1.00)	4.11
3rd (1.31)	5.38
2nd (1.66)	6.82
1st (2.20)	9.04

SPEEDOMETER ERROR

30 mph	actual, 26.4
60 mph	56.6

PERFORMANCE

Top speed (6500), mph	128
best timed run	n.a.
3rd (6500)	98
2nd (6500)	77
1st (6500)	58

FUEL CONSUMPTION

Normal range, mpg	11/17

ACCELERATION

0-30 mph, sec	3.1
0-40	4.2
0-50	5.1
0-60	6.6
0-70	7.5
0-80	9.6
0-100	14.5
Standing ¼ mile	14.2
speed at end	98

TAPLEY DATA

4th, lb/ton @ mph	360 @ 78
3rd	475 @ 72
2nd	580 @ 57
Total drag at 60 mph, lb	135

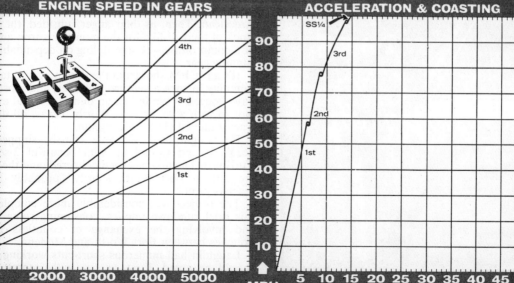

ENGINE SPEED IN GEARS

4th
3rd
2nd
1st

ENGINE SPEED IN RPM

ACCELERATION & COASTING

SS¼
3rd
2nd
1st

MPH

ELAPSED TIME IN SECONDS

ITALIAN ACCENT

*of a Corvette from Fort Worth, Texas,
is furnished by Scaglietti of Modena, Italy*

STORY BY LORIN MC MULLEN

PHOTOS BY AL PANZERA

OWNERS OF EXOTIC CARS who have changed spark plugs in zero weather or broken down on the turnpike have threatened to chuck it all and go straight American. Then, as their anger subsided, they took new stock of their dislike for the monstrous iron and dreamed of a compromise—Italian styling on dependable domestic running gear.

It's an ICBM shot from the dreaming and talking stage to reality in such a venture, however, and most intentions have bogged down before an ounce of metal ever moved. Not so with a young Fort Worth, Texas, oil drilling contractor named Gary B. Laughlin.

He bought three Corvette chassis, shipped them to Scaglietti and wound up with beautiful 2-seater Gran Turismos. The first has been delivered, and the two others are due in Fort Worth shortly (probably before this reaches print).

The project was considerably more complicated than this brief rundown implies, consuming some 18 months and involving the exchange of countless letters and sketches between Fort Worth and Modena.

Laughlin had numerous short-cuts working for him, it should be noted. First, as one with interest in several Chevrolet dealerships, he was granted an audience with top Chevrolet brass toward getting the chassis and engines in various stages of tune.

And, in his work as a drilling contractor, which takes

him to all parts of the world, Laughlin found it convenient to make detours to Modena, home of Scaglietti.

Thus, he was able to veto compromises and to insist upon features he deemed essential to a car that would successfully merge eye appeal in the grand Ferrari manner with American dependability. Among other things, leg room had to be over-size. Laughlin knew that some of his Texas friends just couldn't get into most of the sleek Italian coupes.

Laughlin turned to Scaglietti because he admired the starkness and functional appearance of his craftsmanship in the Ferrari race cars, as opposed to the plushier touches in the coupes of such designers as Farina and Vignale.

Most observers familiar with the designer's problems think Laughlin and/or Scaglietti did very well in concealing the Corvette's rear chassis width. The fenders curve in from their extreme width to a false fender line, which is creased and rolled into a graceful sweep that blends into a top not much wider than that on a genuine Ferrari.

The long, sloping hood, of course, was easy for Scaglietti. Laughlin insisted on only one Corvette touch, and that is in the grille.

The first car is equipped with the 315-bhp fuel injected engine and is otherwise standard Corvette. Because over-all weight is some 400 lb less than a standard Corvette, performance compares favorably with the very best street machines.

The next two cars will be full "slushomatic," Laughlin said, and pointed out that there is ample room for air conditioning. Additional cars could doubtlessly be turned out in far less time than the 18 months required to get the prototype rolling and Laughlin estimates the selling price at about half that of Enzo's incomparable vehicles.

Laughlin comes by his design-engineering interests legitimately. A petroleum engineer and fighter pilot in World War II and the Korean war, he was one of the pioneers in SCCA racing in Texas. He raced nearly everything from the old Allards to Birdcage Maseratis, and still has a 3-liter Ferrari Testa Rossa.

THE 1963 CORVETTE

A Technical Analysis

BY JOHN R. BOND

WITH THE ADVENT of a completely new Corvette (except for the engine and transmission), we are reminded of the fact that there wouldn't have been such a car at all—except for a couple of timely coincidences.

When Harley Earl, General Motors' styling chief (now retired), got the sports car bug, the stage was set. And Chevrolet's chief engineer at the time (Ed Cole) did the rest—he had owned a Jaguar XK-120 and a Cadillac-Allard.

Early in 1952, the GM styling section began work on some special cars for the January 1953 Motorama. On June 2, 1952, Chevrolet engineers were shown a plaster model of a proposed car having a wheelbase of 102 in.—the same as the XK-120. Project "Opel" it was called, and it was up to Chevrolet to provide a chassis and engine so that this show car would at least be driveable.

Project Opel was completed in time, and the 2-seater roadster first shown to the public was renamed the Corvette. During the design stages some consideration of possible produc-

tion was entertained, and the reception of the car was so overwhelming that it was announced 300 examples would be produced. *Road & Track* said, in April 1953, "Chevrolet and Kaiser-Frazer are racing to see which will be first on the market with a volume-produced American sports car. Chevrolet's Corvette will be powered by a souped-up engine [the 6-cyl] delivering 160 bhp. K-F's car, designed by Howard Darrin, will be known as the DKF-161. Both will have fiberglass bodies, at least at first."

Chevrolet won the "race" and kept its 1954 promise—"We are in the sports car business to stay." The prototype had 160 bhp at 5200 rpm but production cars had 150 bhp. Originally, Chevrolet planned to build the first 300 bodies in fiberglass—and then convert to steel. But experience with the new material was so successful that the change was never made.

Until now, no really drastic changes have been made in the overall design, except of course the switch to a V-8 engine late in 1955. The original X-type frame, the 1953/54 passenger car front suspension, the basic body panels, etc.—all

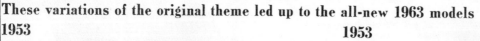

These variations of the original theme led up to the all-new 1963 models

1953   1953   1956

The first Corvette was in the showrooms in late 1953.

Corvette-based Corvair was 1953 Motorama show car.

have remained unchanged except for a few minor details.

But for 1963 we have, as we hinted earlier, a completely new car. First of all, the new car is smaller than before. Secondly, there is a completely new chassis; a new frame, new front and rear suspension—even a new body type, a fastback coupe. In effect, there are three body types for 1963, a folding-top convertible, a hardtop option, and the fastback model. The engine line-up remains the same as last year: a 327-cu-in. engine in 4 options: 250, 300, 340 or 360 bhp.

To give an idea of the size change, the wheelbase is 4.0 in. shorter (now 98.0). The rear track is 2.0 in. narrower and the car's frontal area (by our arbitrary 80% of o.a. dimension formula) is reduced from 20.4 to 19.25 sq ft. Yet all interior seating dimensions are the same, or better, than before. The body material remains fiberglass, but there is an inner network of steel reinforcements around the door openings and under the cowl. (Some aluminum was used before, particularly at certain mounting points.)

Though the smaller car should be (and is) lighter, the new body is heavier so the net weight saving is only about 50 lb.

New body features, in addition to the steel reinforcing structure, include curved side-window glass, a new windshield, cowl-top ventilation, retractable headlights and an improved heater. The luggage compartment has been completely redesigned to give more useful space but there is no lid—access is by tilting the seat backs forward. The spare tire is carried in a sealed fiberglass housing, which is hinged so that its rear end drops down to the ground when released.

The chassis changes have been designed primarily to improve handling qualities. Even the basic weight distribution has been changed. Before, the Corvette carried 53% of its curb weight on the front wheels. The new distribution is 48/52, front to rear. This change, in conjunction with the revised suspension, reduces the former understeer to practically zero.

The front suspension is the familiar ball-type used on Chevrolet sedans since 1955 and it incorporates the anti-dive feature (upper arm pivot tilted about 9°). With this change,

1963

First body change was in 1956.

1959

Dual headlights appeared in 1959.

1961

"Sting Ray" rear end treatment came in 1961.

Steel reinforcing is built into the coupe body.

Electric mechanism for headlight retraction.

1963 CORVETTE

the front cross member is now welded and no longer bolted to the frame. The rear suspension is all new and fully independent. The 3-link geometry is quite similar to that of Formula I racing cars and is clearly shown in the illustrations. Here the double-jointed open driveshaft on each side also serves as a suspension link and, with the control rods, forms the usual short and long arm (s.l.a.) geometry. Trailing radius rods take the brake and propulsion loads.

The rear spring is arranged almost exactly the same as the one used at the front of the 1937 Cord. This is a transverse 9-leaf assembly with rubber cushioned struts in tension at each end to make the connection. Advantages claimed for this all-new suspension system include improved ride and handling, lighter weight, reduced unsprung weight and elimination of rear-wheel tramp. The differential and spring are bolted to the rear cross-member and rubber is used between the member and the frame side rails.

It might be asked—why a leaf spring at the rear? The answer is simply that there is no room for coils except behind the rear axle. This would add both more total and more unsprung weight. Chevrolet points out that only the tips of the leaf spring are unsprung weight. Furthermore, a little study of the photo below and to the right shows that it would be very awkward to fit torsion bars into this layout, arranged either transversely or longitudinally. As it is, the trailing links extend just a little behind the wheel hubs to provide the tension strut mounting.

Along with the improved weight and handling features, we find some very interesting changes in the steering. First, there is a new recirculating-ball steering gear to reduce friction. Then, the center idler arm system has been abandoned in favor of the more accurate dual-arm, 3-link geometry. These

changes, along with the ball-jointed i.f.s., allow faster steering with less effort. There is a very clever change which gives any purchaser a choice of steering ratio. Two tapered holes are provided in each steering arm for the usual ball stud tie rod end. A simple service operation can change the steering from the standard ratio of 19.6:1 (formerly 21.0) to 17.0:1, overall. The turns, lock to lock, are 3.40 and 2.92, respectively, both slightly quicker than in 1962, when a special adapter plate was necessary to get the faster steering feature. There are three other interesting changes in this area: 1) a hydraulic steering damper is used, 2) a power-steering unit of the linkage-booster type is optional (except on the 340 and 360-bhp models) and, 3) a rubber bushed universal joint connects the steering shaft to the gear. This joint has a unique adjustment, arranged so that loosening two bolts will allow the steering wheel to be moved in or out by 1.5 in., or a total movement of 3.0 in.

A new frame was mentioned earlier. Although it has no X-member, it is lighter than before, yet considerably more rigid in torsion. The side rails have been moved farther outboard for better body support. There are 12 body mounts and these tie in with the steel body reinforcements to give added beam strength. An unusual detail which certain other sports car manufacturers could copy is the attention paid to proper ground clearance for the dual exhaust system. The large cross-member under the seat has welded-in tubes which allow routing the exhaust pipes through, rather than over or under.

There are some brake changes, including adoption of the self-adjusting feature. The front drums are wider and the total drum swept area is raised from 259 to 328 sq in. Metallic (iron) linings remain optional and there is a "Special Performance" metallic brake option with larger diameter Al-Fin drums (334.3 sq in., swept), a dual master cylinder and a vacuum booster. These special brakes have a self-adjusting feature that compensates for wear when the car is moving

Coil spring front suspension with ball joints.

Differential bolts to chassis—reduces unsprung weight.

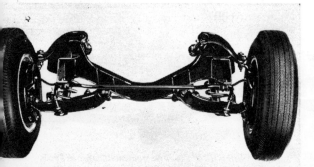

forward—the standard Corvette system requires reversing to adjust for wear. The former extra-wide rim wheel option (15 x 5.50K) is now standard equipment for all 1963 Corvettes.

Tire size remains 6.70-15 and, while most of the 1962 gear ratios are retained, the exception is the two engines with mechanical lifters (340 and 360 bhp). Formerly, these engines came with a 4.11:1 axle ratio; now the standard ratio is 3.70:1.

Engine changes are minor: a Delco alternator replaces the generator and there is closed crankcase ventilation. A smaller flywheel allows the engine to be lower and this requires a new clutch housing, which is aluminum as before. There are minor changes inside the plenum chamber (which is larger, too) of the fuel-injection intake manifold and the f.i. system gives an even quicker throttle response.

A special optional equipment group is listed as available only on the fastback coupe—which shows which body style is destined for real competition. The list includes heavy-duty springs and shock absorbers, a stiffer front stabilizer bar, cast aluminum wheels with genuine knock-off hubs, a 36.5-gal. gas tank, finned aluminum brake drums, power-operated brakes with metallic linings and a dual master cylinder.

One entirely new option for all models is genuine leather seats in place of the normal vinyl-plastic.

Considerable work was also done in the field of aerodynamic research, both at the GM Tech Center and at Cal Tech in Pasadena. One of the results of this study was the decision to use retractable headlights. These are operated by a pair of electric motors and definitely lower the drag factor when closed. In case of trouble, the lights can be rotated into operating position manually by turning cranks which are located under the hood.

The 1963 Corvette has come a long way in 10 years—in fact, from a stylists' plaything to a full-blown, out-and-out dual-purpose sports car. ◈

he only U.S. automotive use of transverse leaf.

New CORVETTE Sting Ray by Chevrolet

ONLY A MAN WITH A HEART OF STONE COULD WITHSTAND TEMPTATION LIKE THIS! The new Corvette Sting Ray is about all the car a red-blooded American-male enthusiast could ask for. The beautifully efficient Sting Ray styling comes in a two-seater sport coupe and a convertible. Both have electrically operated retractable headlamps, completely new interiors with beautifully businesslike instrumentation that'll break your heart, crank-operated ventipanes, and a driving position that's certain to bring out the hero driver in all of us. But if you think the looks are spectacular, take a look underneath! A completely new chassis design that's shorter, with more torsional rigidity, puts engine and driver farther back and results in a rearward weight bias for better handling, greater stability. Link-type independent

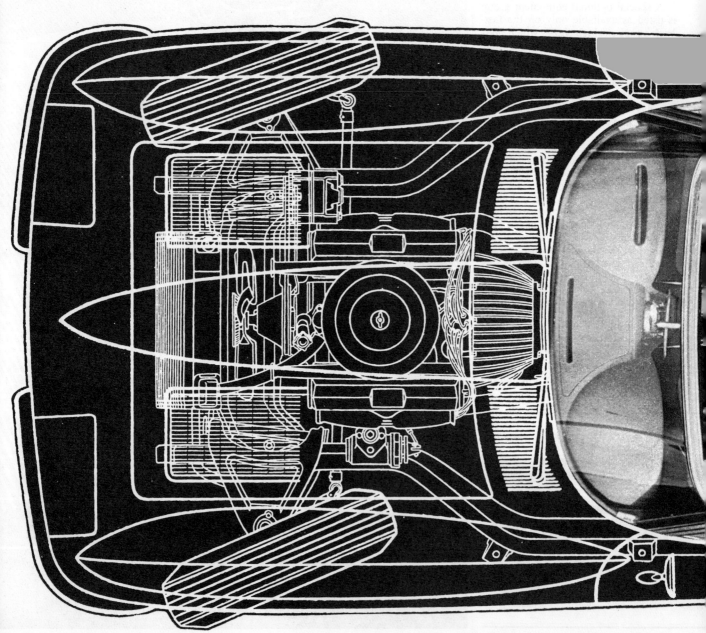

ear suspension keeps rear wheel camber nearly constant from bounce to rebound to get ll that power on the ground. It has new Ball-Race steering with a hydraulic shock bsorber built into the linkage, a built-in provision for quickening the ratio, and a steer-ng column that has three inches of fore and aft adjustment. It has 18% larger self-djusting brakes. The standard wheel rim width has been increased to 5½ inches for etter bite. It has a list of extra-cost options that range all the way from Fuel njection through finned brake drums with metallic linings to knock-off aluminum wheels. The new Corvette Sting Ray is a 100% improvement over the old Corvette, nd we're pretty sure everybody remembers how good *that* was! . . . Chevrolet Division of General Motors, Detroit 2, Michigan. **NEW CORVETTE STING RAY BY CHEVROLET**

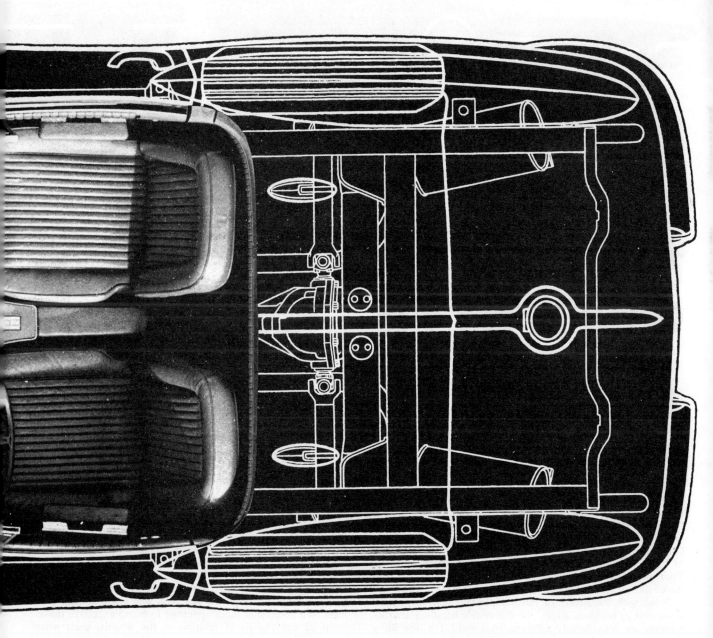

1963 CORVETTE

America's only production sports car is smaller, lighter (with better weight distribution) and has all-independent suspension

NOT SINCE 1960 has *Road & Track* tested a Corvette and, although we gave this devil his due ("The Corvette is unmatched for performance per dollar in terms of transportation machines . . ."), there was a desire expressed by our testers to "do something about the rear suspension." The tendency for the rear wheels to spin freely on acceleration and for the rear end to come sliding around rather quickly during hard cornering was always there. Chevrolet engineers had done a good job with what they had at hand, but there just wasn't enough with which to work. That production-component live rear axle could hop and dance like an Apache with a hot foot.

Now, with the advent of full independent rear suspension described in previous pages, the Corvette's handling characteristics are considerably different. In a word, the new Sting

Ray sticks! Whether you slam the car through an S-bend at 85 or pop the clutch at 5000 rpm at the drag strip, the result is the same—great gripping gobs of traction.

We proved it to our complete satisfaction on General Motors' infamous "Ride and Handling Road." Here, all GM divisions test just those factors in their own, and on competitors', vehicles. The 3-mile route includes samples of virtually every kind of road one could ever encounter. The main loop, while not smooth, is a satisfactory substitute for a road racing course. There are several sweeping right-handers, a keen S-bend and a sharp-right, sharp-left series that (purposely) is made even more interesting by rows of 1-in. ripples running transversely across the roadway. Obviously, if a car is pushed hard through here, a hopping rear axle will lose any semblance of traction; we suspect many a GM product has gone off the pavement, tail first, in this spot. But not so our test Corvette; we entered this with caution the first time through, testing traction with short bursts of throttle. Although we got lots of shaking and juddering throughout the car, it seemed to hold quite well. Second time through, more power, more speed. Third time around we gritted our teeth and held on—5000 rpm in 2nd gear produced just a trace of "scrabbling" at the rear while the whole car remained completely in control. Older Corvettes develop an incurable case of St. Vitus's Dance just at the sight of this corner.

The S-bend was even more fun: Every time through it we discovered we could have gone a little faster. We never did find the limit, although the last pass was made at nearly 90 mph. We noted that there appeared to be no excessive movement of the chassis in relation to the wheels, even when

straightening out fast bends such as these. It does have that distinct "walking" feeling peculiar to virtually all independently sprung cars—one senses the wheels working, moving up and down, one at a time, independent of the supremely stable platform.

One other incident further emphasized the car's greatly increased traction. When we started our acceleration runs we ran the engine up to 3000 rpm and began to ease out the clutch. On the older Corvettes this would have been enough to set the rear wheels a-spinning and the car off to a good, quick start. On the Sting Ray, nothing happened except that the car sort of lurched away from the starting point. We tried it again with 3500 rpm, then 4000 and 5000. The truth was apparent: traction is so good you have to wind up to at least 4500 rpm before you can induce off-the-line wheelspin. And, at that, the wheels don't spin very much. Our test car, despite being a pilot-line model (these are cars run down a preliminary assembly line to test procedures and are therefore not always the mechanical equals of regular production cars) and equipped with a 3.70:1 differential ratio, was faster up to 60 mph than our 1960 test car, which had a 4.10 ratio (but a 283-cu in. engine). As a matter of comparison, here are the acceleration figures for that 1960 f.i. Corvette, our 1963 test car and a 1962 (f.i., 327 cu in., 4.11:1) Corvette tested by *Car Life* magazine:

	1960	1962	1963
0-40	4.2	3.6	3.4
0-60	6.6	5.9	5.9
0-80	9.6	9.3	10.2
0-100	14.5	14.0	16.5
¼-mile	14.2	14.0	14.9

The increased traction gives the '63 equal or better acceleration up to 60 mph, but from there on up the lack of a 4.11 axle ratio hurts its performance in comparison with the others. (As a point of interest, Corvette Positraction gearsets are available in 3.08, 3.36, 3.55, 3.70, 4.11 and 4.56 ratios.) Because the test car was a pre-production model, and the test driving was limited to the GM test track, *Road & Track* intends to test a production model at a later date.

There are many, many improvements elsewhere in the Corvette, including completely new (although production sedan components) front suspension, frame, seats, outside panels, inside panels, dash, steering wheel, *ad gloriam*. When you get right down to it, the car is virtually all new, using mostly only the education gained from the old one.

One thing the designers thought of this time around—the driver. Not only is the steering wheel adjustable for reach (3 in., in and out) so that you can drive Italian-style, but the seats are comfortable and give enough leg room. Instruments are all new and better placed, with twin 6-in. speedometer and tachometer directly in front of the steering wheel. The passenger still has a panic handle in front of him, but it isn't quite as obvious as before. Also, there's a locking glove cubby (gin bin, to the country club set). Our only complaint about the interior was in the coupe, where all we could see in the rear view mirror was that silly bar splitting the rear window down the middle.

While our test car was a fuel-injected, 360-bhp convertible, we also drove the new coupe. Both have one major drawback in common—the lack of easy entry to storage space behind the seat. Both cars have adequate luggage space, but the lug-

Convertible top, and luggage, is stored under cover.

1963 CORVETTE

gage, or anything else that is hauled in that area, must be put in from the front; there are no deck lids. In the case of the convertible, the top must be disconnected from its tonneau panel and the seat backs flipped forward before access is gained. In the coupe, there's room for a couple of young children and lots of baggage behind the seats, but they all have to go in through the doors. (One of those Aston Martin type rear-window doors would have been an ideal solution—perhaps Chevrolet could offer one as an option.)

The coupe is very quiet inside, with little wind noise below 70 mph. Available with power steering and power brakes, and Powerglide automatic transmission, it probably will earn Corvette a bigger share in the burgeoning "personal car" market.

As a purely sporting car, the new Corvette will know few peers on road or track. It has proved, in its "stone-age form," the master of most production-line competitors; in its nice, shiny new concept it ought to be nearly unbeatable.

The headlight buckets are rotated into operating position by electric motors—with stops for correct height adjustment.

New raked-back windshield and removable top are vast improvements over previous models. Design is busy but well integrated.

ROAD TEST
1963 CORVETTE

SCALE: 10" DIVISIONS

DIMENSIONS

Wheelbase, in	98.0
Tread, f and r	56.3/57.0
Over-all length, in	175.3
width	69.6
height	49.8
equivalent vol, cu ft	351
Frontal area, sq ft	19.3
Ground clearance, in	5.0
Steering ratio, o/a	19.6
turns, lock to lock	3.4
turning circle, ft	n.a.
Hip room, front	2 x 20.5
Hip room, rear	n.a.
Pedal to seat back, max	40.8
Floor to ground	7.5

CALCULATED DATA

Lb/hp (test wt)	9.4
Cu ft/ton mile	154
Mph/1000 rpm (4th)	21.8
Engine revs/mile	2750
Piston travel, ft/mile	1490
Rpm @ 2500 ft/min	4615
equivalent mph	101
R&T wear index	41.0

SPECIFICATIONS

List price	n.a.
Curb weight, lb	3030
Test weight	3380
distribution, %	48/52
Tire size	6.70-15
Brake swept area	328
Engine type	V-8, ohv
Bore & stroke	4.0 x 3.25
Displacement, cc	5340
cu in	326.7
Compression ratio	11.25
Bhp @ rpm	360 @ 6000
equivalent mph	131
Torque, lb-ft	352 @ 4000
equivalent mph	87

GEAR RATIOS

4th (1.00)	3.70
3rd (1.31)	4.85
2nd (1.66)	6.14
1st (2.20)	8.14

SPEEDOMETER ERROR

30 mph	actual, 28.5
60 mph	58.0

PERFORMANCE

Top speed (6500), mph	142
best timed run	n.a.
3rd (6500)	108
2nd (6500)	85
1st (6550)	65

FUEL CONSUMPTION

Normal range, mpg	11/14

ACCELERATION

0-30 mph, sec	2.5
0-40	3.4
0-50	4.5
0-60	5.9
0-70	8.0
0-80	10.2
0-100	16.5
Standing ¼ mile	14.9
speed at end	95

TAPLEY DATA

4th, lb/ton @ mph	430 @ 75
3rd	550 @ 68
2nd	off scale
Total drag at 60 mph, lb	120

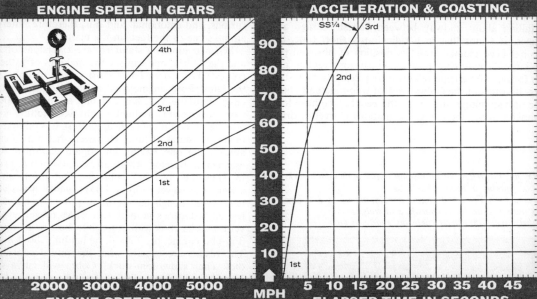

ENGINE SPEED IN GEARS

ENGINE SPEED IN RPM

ACCELERATION & COASTING

ELAPSED TIME IN SECONDS

CORVETTE vs. COBRA: the battle for supremacy

CONTROVERSY is the lifeblood of automobile racing, and the sport has recently been given another of its frequent transfusions. The big battle now being waged is between factions in the AC Cobra and Corvette Sting Ray camps, with the former's shouts having a decided ring of triumph and the latter's falling about midway between honest outrage and sour grapes. It seems that in a ridiculously short time, the Corvette has been clouted from its position of absolute primacy in large-displacement production-category racing, and Corvette fanciers are a trifle reluctant to accept the new state of affairs. Boosters of the Cobra (few of whom actually have any hope of becoming owners) are the people who have long been annoyed to see those big, ostentatious Corvettes drub-

bing the *pur sang* imported sports cars. The undeniable fact that the Cobra is as much *bar sinister* as *pur sang* does not appear to bother this group much; the Cobra looks every inch the traditional hand-built sports car (which it is, to a remarkable degree) and that is enough. In any case, the battle waxes furious, and emotional, and it is, therefore, interesting to examine some of the facts in the matter.

When comparing standard street versions of the Corvette and the Cobra, one can see the makings of rather an uneven contest. The Cobra has a curb weight of only 2020 lb, and the latest Ford engine used as standard in the car, the 289-cu-in. Fairlane V-8, has 271 bhp at an easy 6000 rpm. The Corvette presents a slightly confused picture, insofar as the touring version is concerned, because it is offered with engines in several states of tune. However, that most nearly comparable is the one having an engine equipped with the big, 4-throat carburetor, which gives it 300 bhp to propel its 3030 lb. Thus, the "average" Cobra one finds on the street will have a weight to power ratio of 7.45:1, while its Corvette counterpart, even though having more power, is heavier and has a less advantageous ratio of 10.1:1. Moreover, even if the Corvette purchaser is willing to go "whole-hog," and opt for the 360-bhp engine, he will still be hauling about 8.4 lb per bhp. The results are exactly what theoretical considerations predict. The "showroom-stock" Cobra will cut a standing-start ¼-mile in 13.8 sec, with a terminal speed of 113 mph, while a Corvette, in similar tune, is about a full second slower and will reach not quite 100 mph at the ¼-mile mark.

In top speed, too, the Cobra has the advantage. Its nominal frontal area of 16.6 sq ft gives it quite an edge on the Cor-

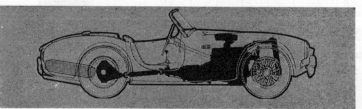

SHELBY AC COBRA: all-aluminum, hand-built coachwork on a chassis having a 90-in. wheelbase and powered by a Ford V-8 engine developing up to 340 bhp. It is less than 152-in. long, overall, and weighs only slightly over 2000 lb at curbside, ready to go.

during cornering, and assume a camber angle that adversely affects cornering power. To compensate, the rear wheels, particularly on the competition Cobras, are given a fairly considerable amount of initial negative camber, so that the "outside" wheel is brought upright as the chassis leans, and that restores much of the tire adhesion that would otherwise be lost. Unfortunately, the tires are cambered too much for the best possible grip under straight-line acceleration. And this is no mere theoretical probability; the competition Cobra is notable for the difficulty it has in getting all of its thunderous horsepower applied to the road surface.

The Corvette Sting Ray, on the other hand, is a very recent design, and incorporates much of what has been proven desirable, in general suspension layout, over the past 3 or 4 years. It has the unequal-length A-arm (with coil springs) front suspension that has, with good reason, become standard for both passenger and racing cars, and a Lotus-inspired unequal-length link rear suspension. The roll centers are at a more modern height than is true of the Cobra, 3.25 in. in front and 7.56 in. at the rear. This, in itself, means that the Corvette will tend to lean a bit less than the Cobra, but the really important thing is that the outside wheels are held in a substantially upright attitude as the chassis leans, and the tires maintain good contact with the road. Also, the front suspension has its members angled upward to provide an anti-dive factor of about 50%, which, of course, cuts nose-dip under braking to half of what it would be without this feature. Finally, somewhat softer springs and longer wheel travel are provided in the Sting Ray's suspension, and the car rides more comfortably than the Cobra—which is, itself, not bad in that respect.

We would say that, in the touring versions, the Cobra and Corvette handle about equally well, with a slight nod in the Cobra's direction because of its lower bulk, weight and quicker steering. However, the Cobra's quick steering, now a rack-and-pinion setup in place of the former cam-and-roller steering box, is not entirely a blessing. The completely reversible nature of the steering box delivers road shocks from the tires right through, undiminished, to the steering wheel, and there are times when cornering hard when wheel-fight can be something of a bother. Here again, the Corvette also has its troubles: its steering, although accurate and free of feed-back, is just a shade too slow, and it is sometimes difficult to wind-on opposite lock fast enough to catch the car's tail as it swings out under a too enthusiastic application of power.

With regard to brakes, the Cobra scores heavily over the Corvette—at least insofar as sheer resistance to fade is concerned. Actually, disc brakes have not yet proven to be as trouble-free in day-in, day-out service as the better drum-type brakes, which the Corvette has.

Taken as touring cars, and bearing in mind all of the factors of reliability, service life, availability of service, comfort, utility, and that most important of intangibles, driving pleasure, it is difficult to make a choice. The Cobra is nominally an import, but the major mechanical elements are American manufactured, and most service problems can be handled by

vette, which is pushing away at 19.3 sq ft of air, and the touring version of the Cobra will exceed 150 mph (urk!), about 10 mph faster than the Corvette—even when the Corvette has the "big" engine. This disparity in top speed will continue, in all likelihood. The airflow over the Cobra is probably not as clean as that over the Sting Ray coupe, but the Cobra's advantage in frontal area cannot be denied. To counter that advantage, the Sting Ray would have to be 14% "cleaner" than the Cobra—and it isn't.

In handling, the two cars are more evenly matched than in any other area. Both cars have all-independent suspensions, and any advantage its lightness might give the Cobra in cornering power is just about offset by its rather primitive suspension layout—the Sting Ray has a much more sophisticated suspension.

The Cobra's basic chassis and suspension were laid down back in 1952, or thereabout, by Tojiero, in England, for a series of very limited production sports/racing cars. These were quite successful, and the design was bought by AC and adopted for its 1954 Ace sports/touring car. The Tojiero design, which borrowed heavily from Cooper's serendipitous Formula III car, has a frame that consists of a pair of large (3-in.) diameter steel tubes, with appropriate cross-bracing, and tall box structures at the chassis ends that carry the suspension elements. These elements are a transverse leaf spring, mounted atop the box structures, with a pair of A-arms underneath, giving an essentially parallelogram geometry and a roll center at ground level. This theme is repeated at both front and rear of the chassis.

With this suspension, the Cobra's wheels tilt with the chassis

CORVETTE STING RAY: futuristic fiberglass panels on a chassis having a 98-in. wheelbase and powered by a Chevrolet V-8 engine developing up to 360 bhp. It is a trifle more than 175-in. long, overall, and weighs slightly over 3000 lb at curbside, ready to go.

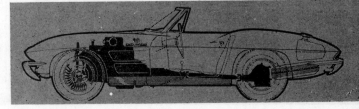

CORVETTE vs. COBRA:

any Ford garage. It is not outstandingly comfortable, if you happen to be talking in terms of driving from New York to Miami, and not about a sporting afternoon on mountainous back-country roads. Conversely, the Cobra is a somewhat more sporty machine on those same twisty roads than the Corvette. As has been said about so many places, the Cobra cockpit is a great place to visit for fun, but you wouldn't want to live there. As for trunk space, there isn't enough in either of the cars under discussion to argue about.

The Cobra's and Corvette's relative suitability as racing cars is seen in their competition records. Their first meeting, at Riverside Raceway last October, was inconclusive, as the Cobra was then only slightly faster than the "prodified" Corvettes running there, and the Cobra took a narrow lead only briefly, to retire immediately with a broken rear stub-axle. Shortly thereafter, the rivals met again, at Riverside once more, and on that occasion the domination of Corvette in its racing category came to an end. Dave MacDonald and Ken Miles, driving Cobras, beat all of the Corvettes (and there were some good ones there) so badly that it was not even a contest. Indeed, just to add insult to injury, Ken Miles made a pit stop after his first lap, ostensibly to have the brakes, or something, inspected, and after all of the Corvettes had gone by, he set out in pursuit. Whittling away at the Corvettes at the rate of about 5 sec per lap, on a 2.6-mile course, Miles caught his teammate, MacDonald, and relegated the first Corvette to 3rd-place in what seemed like no time at all.

The next confrontation was at the Daytona 3-hour, where a vast comedy of errors prevented the Cobras from defeating the GTO Ferraris (even though they demonstrated that they had the necessary speed) and Dick Thompson, in a Sting Ray, beat back the faltering Cobras to one-up them in that race. In the very recent Sebring Enduro, neither the Cobras nor the Corvettes fared particularly well. A rash of broken engines, and one transmission, eliminated 4 of the 7 Corvettes entered, and one of those still running at the race's end had been in the pits for a majority of the 12 hours having its engine bearings replaced. This Corvette completed only 46 laps.

The showing down at the snake (Cobra) pit was a little, but not much, more impressive; they lost exactly half of the 6 cars entered, and all of the finishing Cobras had to be nursed back from the ranks of the walking wounded at least once during the race. Even so, the Cobras' showing was better than the results indicate. Most of their problems were of a relatively minor nature (no shattered engines or other major components, at any rate), and while they were out on the course the Cobras showed more sheer speed than almost anything there. Phil Hill was observed, in practice, engaging one of the "prototype" Ferraris in a drag race up the pit straight and the good Phil, smiling hugely and rowing away at the gear-lever, carried it to a draw going into the first turn—after

which the Ferrari moved away in no uncertain fashion. The Cobras, while they were in action, had plenty of speed, and the best-placed Corvette finished 10 laps behind the first of the Cobras.

One of the more interesting aspects of the great Cobra-Corvette debate is that the "Chevrolet-Forever" contingent has been complaining bitterly about the "unfair advantage" Shelby has taken in securing a list of approved competition options for his Cobras. This is indeed curious, for the ploy under attack is precisely the one used by GM to make its Corvettes competitive. In fact, we can draw parallels between almost every option offered for both cars. The Corvette has its fuel injection; the Cobra a double brace of 48-mm, double-throat downdraft Weber carburetors. Both have optional competition brakes with friction material largely unsuited to street-type driving. Aluminum alloy, cross-flow radiators are offered for both, as are competition exhaust systems, and cast light-alloy wheels can, due to a relaxing of the production car racing rules this year, be used on any car. Special, and very stiff, springs are catalogued for each car, as are dampers, and there are the miscellaneous items like oversized fuel tanks, for distance events, and more axle ratios than anyone could hope to need for either car. Transmission ratios? They are identical, each car using the same Warner Gear transmission. The Corvette is delivered with the close ratio gears for this gearcase installed as standard, and the wide ratio gears are offered as an option; the Cobra comes standard with wide ratio gears and the close ratio set is available as an option.

In full racing trim, both the Cobra and the Corvette would be thoroughly unpleasant to drive down to the office. The hot-cam, fuel-injected Corvette engine rumbles and chuffs smoke at low speeds, and so does the 340-bhp (at 6500 rpm), Weber-carbureted racing engine in the Cobra. Clutch and brake pedal pressures in both cars are fierce, and the low-end throttle response is awful. The major sin of the Cobra, in the Corvette booster's eyes, is that it is a winner, and it is likely to stay one unless a lightweight version of the Corvette is introduced. These cars' merits as touring machines can be argued, but there is no disputing which is the better racing car. The Cobra's lightness allows it to accelerate and corner faster, and stop quicker (primarily due to the advantage provided by its disc brakes), and on a straightaway of a likely length, the Cobra will be a good 10-mph faster. Given those points, it is very hard to imagine that any well prepared, well driven Cobra will be beaten this year—not by the Corvettes, and possibly not by anyone, unless the organizers get sneaky and push the Cobras over into the same races with all-out racing cars. There are, as a matter of fact, rumors of this happening, and if it does, the Cobras just might beat the big modified cars, too.

No matter where they run, the spectators will be the winners, for the Cobra is fast, noisy and slides about in a spectacular manner, and everyone will eventually learn to admire it for the tremendous sporting/racing machine it is—even the people who drive up to the spectator gate in a Corvette. ◉

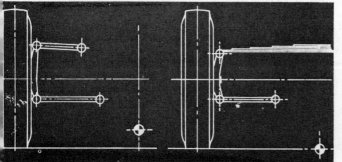

The Corvette front suspension, left, is more modern and has a higher roll-center than the Cobra's, at right.

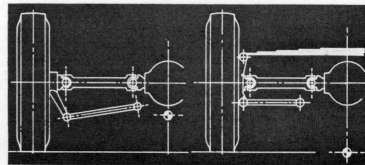

The Corvette's rear suspension has a higher roll-center than that of the Cobra, which as at the front, uses its spring as a link.

FOR THE NEW READER

An explanation of the method of obtaining, and the meaning of, recorded data in Road & Track's road test data panels

ONE OF THE problems created by *Road & Track* having been established for so many years is a tendency for our terms and references to get a trifle "in-group," which is fine for the bulk of our readers, who have been with us a long time, but apt to be confusing for the new reader. Hence, it is periodically necessary to bring our new readers up to date, with a kind of familiarization course relating to automotive terminology in general and the terms used in our data panels in particular.

Many of the items listed in the data panel are, to a great extent, self-explanatory. Even so, there are things which offer a maximum opportunity for

DIMENSIONS

Wheelbase, in	102.3
Tread, f and r	58.4/58.2
Over-all length, in	187.8
width	74.0
height	53.5
equivalent vol, cu ft	429
Frontal area, sq ft	21.6
Ground clearance, in	5.0
Steering ratio, o/a	n.a.
turns, lock to lock	2.8
turning circle, ft	33
Hip room, front	2 x 23
Hip room, rear	n.a.
Pedal to seat back, max	39.0
Floor to ground	13.2

misunderstanding. Under *dimensions*, *equivalent volume* may look mysterious, but it is simply length x width x height. This figure has only a very rough relationship to the true volume (as would be found by submerging the car in a gigantic graduated beaker), but it is an adequate index to the car's bulk for comparative purposes.

Just below equivalent volume, there is *frontal area* and, although the term itself is easily understood, the means by which we obtain the figure is obscure. To get the exact frontal area, it would be necessary to make an accurate front-view drawing of the car, and then laboriously calculate the area in bits, including the separate areas of things like door handles, outside mirrors, low-hanging oil sumps, etc. Given unlimited time, and a huge technical staff, this could be done; in practice, such extreme measures are not necessary. We simply multiply width and height, and then take 80% of the resulting area. Experience indicates that this method produces surprisingly accurate results, and it has the prime virtue of being quick.

The last items listed under dimensions that might require explanation are *steering ratio, overall; turning circle* and *hip room. Steering ratio*, as we give it, is the ratio between rotation at the steering wheel and the resulting motion down at the steering spindle. A bicycle, for example, has a steering ratio of 1:1. *Turning circle* is not the circle described by any point on the centerline of the car; it is, rather, the circle made by the "outside" front wheel, which means that we are using what is called in engineering circles, a "between curbs" figure.

Under *hip room, front*, we occasionally will give a dimension prefixed by 2 x. This is done in cases where the front seats are separate (bucket, as opposed to bench, seating) and the figure given will be for the width of one seat, with the 2 x to indicate that 2 seats are provided. In rare instances, such as a GT coupe with 4 individual seats, the same system would be used for both front and rear seating.

Specifications is a category that is like dimensions to the extent that it is

SPECIFICATIONS

List price	n.a.
Curb weight, lb	3550
Test weight	3850
distribution, %	52/48
Tire size	6.50-16
Brake swept area	446
Engine type	V-12, ohc
Bore & stroke	3.46 x 2.68
Displacement, cc	4962
cu in	302.7
Compression ratio	8.50
Bhp @ rpm	360 @ 7000
equivalent mph	170
Torque, lb-ft	311 @ 5000
equivalent mph	121

mostly self-explanatory. There are exceptions, though: *curb weight* and *test weight* are not defined. The former is the weight of an automobile carrying a full supply of water, oil and fuel, and any tools (such as tire-changing gear) that come as standard equipment. The latter, *test weight*, is the actual weight of the car during our performance tests, with test equipment and our 2-member crew. In almost every instance, the test weight will include all of the items that go into the curb weight, but we give the car whatever benefit may come from a slight reduction in weight by doing the tests with only a half-tank of fuel.

Brake swept area indicates the total area of metal swept (rubbed) by the friction pads or linings. At one time, we used lining area, but with the advances in drum brake design, improvements in lining materials (such as the sintered-metal linings) and the advent of the disc brake, swept area became a more accurate reflection of overall braking capabilities. In fact, with the development of so many different grades of lining material, lining area no longer even gives a valid indication of the brake's probable service life.

Displacement is a term describing engine size. To get total displacement we take the volume displaced by a complete stroke of one piston, and multiply it by the total number of cylinders. We give the displacement in both cubic inches and cubic centimeters, because the engine sizes for American cars are customarily given in the former and imports in the latter, and by presenting both we make it easier for the reader to compare all engines. Also, there is a strong likelihood that the new reader will not have had previous experience with the metric system.

To someone who is completely unfamiliar with the piston-type, internal combustion engine, the piston-displacement sizing system may not, at first, make much sense, but an engine of the type just described derives its power from the expansion of air heated in the burning of a hydrocarbon fuel (the oxygen necessary for combustion being borrowed from the air that is the working gas) and because piston displacement reflects, rather accurately, the quantity of air that will be pumped through the engine, then displacement is a good index of probable power output, and a peerless predictor of torque.

Compression ratio is the ratio between the volumes of the cylinder and combustion chamber with the piston at the bottom, and then at the top, of its stroke. As we have explained (very briefly), the expansion of heated air is the internal combustion engine's source of power, and it should be obvious that by compressing the air into a smaller volume before heating it, there will be relatively more expansion during the piston's power stroke and more power will be obtained. Also, for a variety of reasons, the engine's efficiency will be improved.

The upper limit for compression ratio is determined primarily by the fuel used. At some level of compression, gross combustion irregularities (usually pinging) occur and an engine's compression ratio must, therefore, be held to whatever level available fuels will permit. For the Indianapolis racing engine, where regulations and racing-type budgets allow the use of alcohol fuels, compression ratios may be, say, 14:1; in an economy-oriented sedan, where the manufacturer anticipates the burning of standard, or even sub-standard grades of gasoline, the compression ratio may have to be held down around 7:1.

Bhp is, of course, brake horsepower, a unit of work that was established by the immortal James Watt as 33,000 foot-pounds per minute. This is the amount of work needed to lift a one-pound weight 33,000 feet in one minute; or 33,000 pounds one foot in one minute, etc.

Torque is the twisting force, measured in pounds-feet, developed by the engine. Imagine, if you will, a shaft with a lever extending out of its side, and then further imagine a one-pound weight hanging from that lever at a point exactly 12 inches from the centerline of the shaft. This setup would produce one pound-foot of torque at the shaft. Torque, unlike brake horsepower, is more or less independent of engine speed, and is most useful in propelling a car about in traffic. If we were to depend on horsepower to move a car, it would be necessary to change gears frequently to keep the engine operating

CALCULATED DATA

Lb/hp (test wt)	10.7
Cu ft/ton mile	113
Mph/1000 rpm (4th)	24.2
Engine revs/mile	2480
Piston travel, ft/mile	1105
Rpm @ 2500 ft/min	5610
equivalent mph	136
R&T wear index	27.4

up in the speed range where power is developed; torque reaches a maximum value, in most engines, at a fairly low speed and stays at a high level over a very wide range, making it of great im-

portance in a touring-type engine.

Our *calculated data* section looks formidable, but there is really nothing very complicated contained therein. Pounds/horsepower (*lb/hp*) is simply the number of pounds of vehicle weight per horsepower. This provides a good index of performance potential; a 100-bhp engine will produce much better acceleration in a 1500-lb car than in one that weighs 3000 lb.

Cubic feet per ton mile (*cu ft/ton mile*) is another performance index. The term refers to the working volume (in cubic feet) of air pumped through the engine per mile, per ton of vehicle weight. In a 4-stroke engine the figure is calculated by dividing the displacement in half (to take into account the fact that only half of the pistons' down strokes are used for power; for 2-stroke engines, the full displacement is used) and then multiplying by the number of engine revolutions per mile. This produces the cubic feet per mile; and then we divide by the car's weight, in tons, to get the final figure.

Mph/1000 rpm means, simply, the number of miles per hour of road speed the car will have for every 1000 revolutions per minute of engine speed. If the mph/1000 is 15, then at 15 mph the car's engine would be turning 1000 rpm. At 30 mph the engine speed would be 2000 rpm; at 60 mph the engine speed would be 4000 rpm, etc.

Engine revs/mile is simply what it says: the number of turns made by the engine for every mile the car travels. It is a figure that is valuable mostly because it gives an indication of the number of engine operating cycles per mile, and

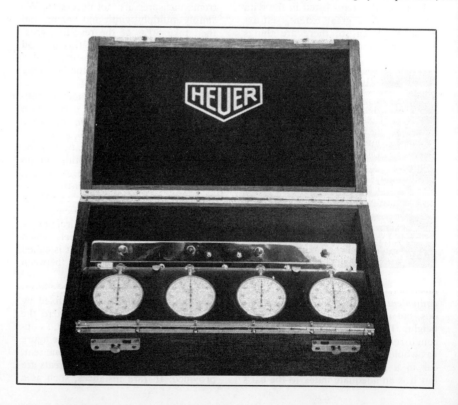

FOR THE NEW READER

is therefore valuable, albeit to a limited extent, in predicting reliability and probable engine life. *Piston travel*, given in feet per mile, is calculated by doubling the stroke (because the pistons make 2 strokes per revolution), multiplying by the number of engine revs per mile and converting the results from inches (the stroke is usually given in inches) to feet. Piston travel, like engine revs/mile, is given because it is an indicator of probable reliability.

The same may be said of the *rpm @ 2500 ft/min* figure, which refers to mean piston speed. Experience, gained with many different engines over the years, has shown that at piston speeds up to 2500 rpm (approximately) inertia loadings will not be high enough to be troublesome. Some engines, of course, have shown a capability for running fairly reliably at piston speeds up to about 4000 ft/min, but this sort of performance is mostly in the realm of the racing engine. Actually, the best figure for us to use would be piston acceleration, but the formula for this requires that we know the connecting rod length, and as this is not always (or even often) available, we substitute the less accurate but more easily obtained piston speed. The limit that is usually given for piston acceleration is, for those who care, 100,000 ft/sec/sec.

The *R&T wear index* is derived by multiplying the engine revs/mile by the piston travel. The resulting figure is too large for convenience so we divide by 100,000. Our wear index figure is open to attack on the grounds that it is arbitrary (which it is), that it ignores differences in engine construction details and quality of material (which it does) and that it also ignores the fact that wear is not entirely a function of the

GEAR RATIOS	
4th (1.00)	3.44
3rd (1.24)	4.27
2nd (1.72)	5.93
1st (2.45)	8.44

total number of engine operating cycles (entirely true). However, despite the many theoretical objections, our experiences indicate that the index is valid, as cars having a high index usually do wear out sooner than those with a low index. The method may be wrong; but results seem to be right and until we can develop a better system, we will continue as we have.

Gear ratios are given both overall (at the right in the data panel) and as

PERFORMANCE	
Top speed (6800), mph	165
best timed run	n.a.
3rd (6800)	132
2nd (7000)	99
1st (7000)	69
FUEL CONSUMPTION	
Normal range, mpg	9/12

transmission ratios (in parentheses). Usually, top gear in the transmission will be direct, and the ring-and-pinion ratio will be given at the upper right. Sometimes, as in the case of the Volks-

wagen, there will be no direct drive and the "true" axle ratio will have to be calculated. Where the transmission ratio given is less than 1.0, the gear is an overdrive.

Under *performance*, we give *top speed* and then a series of figures for the speeds (at some given rpm) in the gears. At one time, we actually timed all of our test cars over a measured distance to get their top speeds, but with increasing difficulty in obtaining sites for this kind of activity, and due to the very high-speed potential of some of the cars we test, we have suspended that phase of the test. In some instances, with some cars, we still do the "flying-quarter," but in many cases we estimate the top speed, using the car's frontal area, power output and axle ratio as determining factors.

Fuel consumption is given over a range that would be normal for the car in average use, which would include some miles spent on a fast trip, some in medium-speed suburban driving and some in traffic. Obviously, driver habit will have some effect, as will one-sided driving conditions, and some owners

SPEEDOMETER ERROR	
30 mph	actual, 25.0
60 mph	51.0

will surely find that their mileage does not agree with that given in our data panel.

Speedometer error is given only at 30 and 60 mph; but we actually determine the error for all speeds up to 100 mph, or to the car's top speed if that is less than 100 mph. Our system is not

very involved: we time the car, running at a steady indicated speed, over a measured distance. This gives us the true speed, and by repeating this process at 10-mph increments, we can establish a pattern of error over the entire speed range. This error is plotted as a curve, which would show and enable us to compensate for any experimental error on our part.

Acceleration figures need no explanation; we give the speed, and the

ACCELERATION

0-30 mph, sec.	3.5
0-40	4.4
0-50	5.3
0-60	6.6
0-70	8.5
0-80	10.0
0-100	14.5
Standing ¼ mile	14.6
speed at end	101

time (in seconds) needed to reach the speed, from a standing start. The times given are averages taken from several runs, and they reflect the kind of performance one may expect when the car is not completely brutalized, but nothing is really spared, either. Starts are made with either the clutch or tires being slipped (to prevent the engine from bogging down) and all shifts are forced—even if there is some complaint from the synchro mechanism as a result. In short, everything is done to insure that, without taking a chance of breaking the car, the best performance possible will be obtained.

Obviously, there are differences between cars, even of the same model, and we cannot always be sure that our test cars are not atypical. However, we

know approximately what kind of performance to expect and when a car differs too much from our expectations, we double-check to make certain that the example given us for test is fairly representative of the type.

The last item, *Tapley data,* has been

TAPLEY DATA

4th, maximum gradient, %	16.2
3rd	26.4
2nd	off scale
Total drag at 60 mph, lb...	

a major cause of confusion, and we are going to do something about that. The Tapley meter is a British-made accelerometer, giving readings in pounds per ton. The meter also gives grade-climbing ability and, effective with this issue, the lb/ton figures that have been listed in all previous data panels will be replaced by maximum gradient figures, given in percentages. A 100% grade, incidentally, is one in which the road rises one foot for every foot forward; a 50% grade would be one that had a 6-in. use for each foot forward.

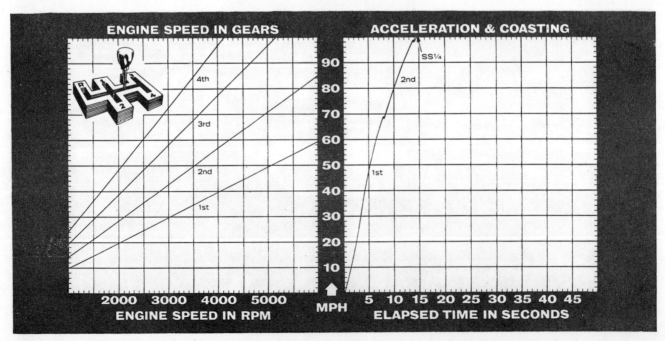

Sting Ray Automatic

Docile but no fossil, and
agile but not fragile

 REFERENCE TO THE Sting Ray usually brings to mind visions of a 375-bhp engine, fuel injection, a 4-speed transmission, and all the accoutrements of a competition sports car. However, this car can be obtained in much more docile forms which make it an ideal fast tourer for those who are not interested in the ultimate in performance. With this in mind, we selected our test car with the 300-bhp version of Chevrolet's 327 engine coupled to an automatic transmission, and found that this "power team," as Detroit puts it, was remarkably lively even if some of the keen edge of the really hot version was lost.

It is 10 years since the Corvette was first introduced, with the Sting Ray model following on in 1963, and the car has built up a remarkably enthusiastic following in that time— even comparable to that of the Porsche. In fact, we were surprised to find that some Corvette drivers still wave to each other when passing, even if they do have a tendency to "whomp" their engines at traffic signals. However, it is definitely a most impressive car in many ways and seems to command a lot of attention and respect on the road.

The body design of the 1964 Sting Ray remains little changed from 1963, although we were pleased to note that the central division in the rear window has been eliminated, with a consequent improvement in vision. The basic lines of the car are most original and extremely clean, but the hand of the stylist has been laid on it at some time so that a rather cluttered appearance results at some angles. Because the car is a "Detroit" product (Yes, we know it's assembled in St. Louis), one suspects immediately that the disappearing headlights are some sort of a sales gimmick, but they definitely do improve the aerodynamics of the car and also help to clean up the general appearance of the front end when retracted.

There was some criticism of the quality of the fiberglass body when the Sting Ray first appeared, but we were unable

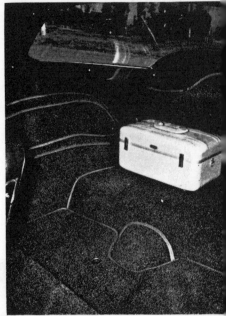

to find any ripples or other faults on our test car. Furthermore, the other characteristics normally associated with fiberglass, such as a tendency toward drumming and magnification of noise, were not present at all, although an occasional creak was audible from somewhere around the rear window when running on uneven ground.

The interior is very pleasing and well finished, and is laid out in the best sports car tradition. Bucket seats are provided for driver and passenger and there is considerable luggage space behind the seats. However, access to this space is through the doors alone, which is extremely awkward when bulky items are concerned, and if the rear window could be hinged in Aston Martin style it would be a great improvement. Another criticism concerns the space allotted to the driver, which is surprisingly small when one considers the overall dimensions of the car.

Applying the standards we used in a recent article on the seating package in various sports cars ("The 99th Man," Jan. 1963) the Sting Ray didn't measure up well at all. Compared to the "average" sports car, the Sting Ray scored 68% and 58% (by the standard for the average and big man), while the average was 80% and 57%. However, provision is made for adjusting the steering wheel, although this cannot be done from the driving seat because it requires the application of a wrench to the steering column under the hood.

In keeping with the other automatic features, the car was equipped with power steering, which we believe to be almost unnecessary on a Sting Ray, but was one of the best we have ever encountered. It was not noticeable at speed, but gave considerable assistance when parking. With an overall ratio of 17.6:1 and 2.92 turns lock-to-lock, the steering was fast, responsive and effortless, and well matched to the diameter of the wood-rimmed, 3-spoke steering wheel.

One of the most interesting features of the car is the independent rear suspension, which is a very successful layout and also a big step forward for General Motors. The system is unusual because it employs a transverse leaf spring, but this is apparently a matter of convenience because there is no room for coils. The two open double-jointed driveshafts make up the top member, and of course these are not splined. Two parallel lower arms pivot low down at the hub carriers and at the differential, and braking torque is transferred to the frame by radius arms.

The result leaves little to be desired because there are none of those qualities, such as a tendency toward rear wheel steering, which one normally associates with independent rear suspension. However, one is aware that the car does not have a solid axle because a slight walking motion is evident under certain conditions of hard driving, but it is never sufficient to be disconcerting and the handling of the car remains completely predictable.

Sting Ray Automatic

AT A GLANCE...

Price as tested	$5016
Engine, cyl/cc/bhp	V-8/5356/300
Curb weight, lb.	3050
Top speed, mph	130
Acceleration, 0-60 mph, sec.	8.0
Passing test, 50-70 mph, sec.	6.0
Overall fuel consumption, mpg	14.8

Sting Ray Automatic

The suspension as a whole is stiffer than we expected but not uncomfortable and does a good job on very bad road surfaces. It is helped considerably by a weight distribution which has a slight rearward bias and this weight distribution has also done a lot for the steering, to the extent that we feel it makes power steering unnecessary under most circumstances.

On the road, the Sting Ray has a giant's stride due to its 300 bhp and 360 ft/lbs of torque at 3200 rpm. The automatic transmission is the Powerglide which, with its two speeds, is certainly not among the most effective transmissions on the market. However, this shortcoming is not particularly noticeable in the Sting Ray because there is power to spare throughout the speed range and the acceleration is exceptional whether the transmission is in high or low.

As far as acceleration is concerned, we were able to draw an interesting comparison with a stick shift Pontiac GTO which we sampled at the same time (see page 32). The Sting Ray tended to squat down on its rear suspension when leaving the line, with never so much as a chirp from its tires, and then gobble up the strip in 15.2 seconds after shifting itself smoothly into high range at 56 mph. In complete contrast, the GTO didn't want to leave the line at all, preferring to sit there burning rubber due to poor weight distribution and lack of an up-to-date rear suspension.

To match the performance of the car, the brakes are adequate for normal fast driving but they will definitely fade and become uneven when used to the limit. When one considers both the weight and speed of the Sting Ray, it would appear to be an excellent car for a disc brake system, and it is surprising that General Motors has not yet adopted discs for this model.

By offering an automatic version of the Sting Ray, General Motors has considerably increased its market for the car. This version is definitely not for the purists, but it is an excellent compromise for those families in which the little woman does her share of the driving, and it has decided advantages for anyone who does the majority of his driving in heavy traffic. With a price tag closer to $5000 than $4000, it is by no means cheap but, on the other hand, it still represents remarkably good value for money when one considers the performance combined with comfort, and the generally high standard of quality throughout the car.

ROAD TEST
R&T Sting Ray

SCALE: 10" DIVISIONS

PRICE

List, FOB St. Louis........$4252
As tested, West Coast......$5016

ENGINE

Engine, no. cyl, type........V-8
Bore x stroke, in......4.00 x 3.25
Displacement, cc..........5356
   Equivalent cu in..........326.7
Compression ratio..........10.5
Bhp @ rpm.........300 @ 5000
   Equivalent mph.............118
Torque @ rpm, lb-ft...360 @ 3200
   Equivalent mph.............75
Carburetor, no., make....1 Carter
   No. barrels..............4
   Diameter:.......primary, 1.56
       secondary, 1.69
Type fuel required.....Premium

DRIVE TRAIN

Automatic transmission:
  2-speed Powerglide
Gear ratios: 2nd (1.00)......3.36
  1st (1.76)...............5.91
  1st (1.76 x 2.10)..........12.4
Differential ratio..........3.36:1
Optional ratios..........6 ratios
      from 3.08 to 4.56:1

CHASSIS & SUSPENSION

Frame type: Full length box-
  section ladder
Brake type.................drum
  Swept area, sq in..........328
Tire size...........6.70x15
  Wheel revs/mi............760
Steering type....recirculating ball
  Overall ratio.............17.6
  Turns, lock to lock..........2.9
  Turning circle, ft.........39.4
Front suspension: Independent
  with coil springs, stabilizer bar.
Rear suspension: Independent with
  fixed differential, lateral leaf
  spring & struts, U-jointed drive
  shafts, tube shocks.

ACCOMMODATION

Normal capacity, persons........2
Hip room, front, in.......2x19.5
Head room, front.............38
Seat back adjustment, deg.....0
Entrance height, in.........34
Step-over height.........15.5
Floor height.................8
Door width.................31.5

GENERAL

Curb weight, lb.........3050
Weight distribution
  with driver, %..........48/52
Wheelbase, in.............98.0
Track, front/rear.......56.3/57.0
Overall length, in.........175.3
  Width.................69.6
  Height.................49.8
Frontal area, sq. ft.........19.3
Ground clearance, in..........5.0
Overhang, front..........32.0
  Rear.................45.3
Departure angle, no load, deg...17
Usable trunk space, cu. ft.....10.5
Fuel tank capacity, gal.......20.0

INSTRUMENTATION

Instruments: 160-mph speedometer, 7000-rpm tachometer, fuel
gauge, ammeter, oil-pressure
gauge, water temperature indicator, clock.
Warning lamps: parking brake,
lights, high beam, turn indicator.

MISCELLANEOUS

Body styles available: Coupe (as
tested), convertible and removable hardtop convertible.

EQUIPMENT

Included in list price: 250-bhp
engine, 3-speed transmission,
tachometer.
Available at extra cost: 300-bhp
engine, automatic transmission,
limited-slip differential, power
steering, power brakes, radio.
(Note: many other options available
too numerous to list.)

CALCULATED DATA

Lb/hp (test wt)............11.3
Cu ft/ton mi..............142.5
Mph/1000 rpm (4th).......23.5
Engine revs/mi...........2555
Piston travel, ft/mi.........1385
Rpm @ 2500 ft/min.......4612
  Equivalent mph...........108
R&T wear index............35.4

MAINTENANCE

Crankcase capacity, qt.........4
  Change interval, mi.......6000
Oil filter type.............paper
  Change interval, mi.......6000
Lubrication interval, mi.....12,000
Tire pressures, f/r, psi....24/24

ROAD TEST RESULTS

ACCELERATION

0-30 mph, sec................3.2
0-40 mph......................4.4
0-50 mph......................6.1
0-60 mph......................8.0
0-70 mph.....................10.5
0-80 mph.....................13.2
0-100 mph....................20.2
Passing test, 50-70 mph......6.0
Standing ¼ mi, sec..........15.2
  Speed at end, mph.........85

BRAKE TESTS

Max deceleration, ft/sec/sec ... 20
  2nd stop..................20

FUEL CONSUMPTION

Normal range, mpg........12–16
Cruising range, mi......240–320

TOP SPEEDS

High gear (2nd),......... mph 130
1st.........................74

GRADABILITY

(Tapley data)

High gear, max gradient, %....22
  1st gear..............off scale
  Total drag at 60 mph, lb....120

SPEEDOMETER ERROR

30 mph indicated......actual 30.0
40 mph....................40.0
60 mph....................59.7
80 mph....................79.0
100 mph...................98.5

ACCELERATION & COASTING

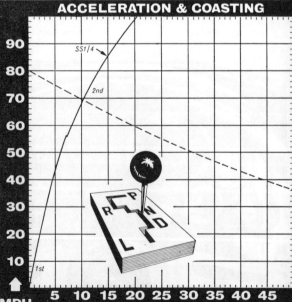

SS1/4

2nd

1st

MPH

| 5 | 10 | 15 | 20 | 25 | 30 | 35 | 40 | 45 |

ELAPSED TIME IN SECONDS

CORVETTE STING RAY

CONTINUED FROM PAGE 77

I have not doted on this car, but I have maintained it well. At the end of the 36,000 miles it was performing almost as new. Obviously it is a long-life car, and one requiring relatively little attention. The higher-powered versions of the Corvette require considerably more attention and get much poorer fuel economy, so we're examining the Corvette that's the most economical to run.

The drum brakes of the early Sting Rays are both good and bad. Good, in that at 36,000 miles the linings are barely half-worn. Bad, in that they often pull unevenly, are able to achieve only 62%-g deceleration, and fade rapidly with hard use.

One of the car's weakest features, I have concluded, is one of the features that attracted me to it: its body construction. It is rustproof, to be sure, but it also has a great propensity for rattles, squeaks and general structural shake on rough roads. De-rattling has become part of the Saturday wash-polish ritual. To balance out my evaluation of the body, I must say that the trim, exterior and interior, was fitted properly—and that the Corvette's convertible top is one of the best. It goes up and down very easily, seals well and wears well too. From the looks of it at this point, I'd say it will last the life of the car.

Another strong point is the AM/FM radio. This is the only radio that can be ordered in the Corvette, and it's not cheap. But it has excellent tone quality and reception and has made many a mile of freeway driving tolerable when it would have been otherwise sleep-inducing. A low noise level makes listening possible at speeds up to 80 mph with the top up.

CORVETTE STING RAY

Overall Cost per Mile for 36,000 miles

Delivered price	$4100
Gasoline	660
Oil & "service station" bills	80
Tires	277
Maintenance & repairs	121
Licensing, 19 mo. @ 12.70/year	20
2 mo. @ 56.00/year	9
Insurance	466
Washing & polishing	30
Total expenditure, 36,000 miles	$5763
Retail value at end of period	3000
Cost of driving 36,000 miles	2763
Overall cost per mile	7.68¢

Even with its smallest engine the Corvette can be quite exhilarating to drive; there's always a great reserve of torque and roadholding on tap, so this is no dullard in any sense. It does serve to show that the car can be owned and operated for a very reasonable outlay. Some of the qualities that attracted me to it are superfluous in the mild California climate where I now live, but I suspect I'll put on at least another 36,000 before replacing it.

1965 CORVETTE STING RAY

*It has the performance,
polish and pizzazz
to suit almost any situation*

TEN YEARS AGO, who could have guessed that the 1965 Corvette would have fuel injection, 4-speed manual transmission, limited slip differential, all independent suspension and, wonder of wonders, disc brakes? The Corvette has become a car that any manufacturer would be proud to produce, and a far, far cry from the 6-cyl phut-phut with 2-speed automatic transmission that was standard in the first model to bear the Corvette name.

The transformation of the Corvette has come about through modest but worthwhile changes that have continuously improved the breed without rendering each previous model obsolete. For 1965, this tradition of development has been continued. The basic selection of body styles—coupe, hardtop and soft top—is unchanged, as is the "Sting Ray" look that was introduced year before last. This year's "identity" changes include the abandonment of the phony louver blocks in the

hood, which we consider a definite improvement, and the replacement of the artificial vents behind the front wheels by three genuine openings. There is also new side trim, different wheel covers, and on the inside there are detail changes around the instruments, seats and door liners. A note in the catalog points out that genuine leather upholstery is also available in the same 10 glorious color-keyed interiors as the standard vinyl.

The big news for 1965, though, is that the Corvette now has disc brakes. Discs are used on all four wheels and are standard on all Corvettes, though you can still obtain drum brakes as a "credit option." These new brakes are named "Roto," no doubt from the rotor effect suggested by the large (11.75-in. dia.), thick (1.25-in.), ventilated disc. The disc starts from a dished out center and then radial fins, which create passages that are open at either end, in effect pump air out from between the two solid flat surfaces wiped by the brake pads on either side. Holes in the splash shields, the wheels and the wheel covers assure ventilation. The brake pads, activated by two pistons on either side of the disc, are large (5.96 x 2.21 x 0.41 in.), which should assure good longevity under normal conditions. The brake swept area is 461 sq in. (compared with the 328 sq in. of last year's drum brakes), which figures out to 7.2-lb per swept sq in. of braking surface; very good indeed.

The action of the foot pedal is progressive in the Corvette system. That is, the amount of braking at the wheels is directly proportional to the pressure applied to the pedal. This is the

best kind as it gives brake "feel" to the driver. The brakes are self-adjusting except that the parking brake, a 6.5 x 1.25 drum machined into the dished portion of the rear hubs, is manually adjusted by the conventional "star wheel" system.

But before going into the effectiveness of these brakes—and they are highly effective—let's take a little longer look at the Corvette itself.

The body, as everyone must know, is fiberglass. A ladder type frame is used, with box section tubes down either side and a total of five cross members.

The suspension of the Corvette is atypical among American production cars, as it is independent at both ends. The front suspension is conventional Detroit with double A-arms, but at the rear the differential is attached to the frame and the power is delivered through U-jointed driveshafts to the wheels, which are located by transverse multi-leaf spring and lateral struts.

In overall appearance, the Corvette is somewhat deceiving because, under normal driving conditions, it seems small, handy and agile. The proportions are good, it seems to us, everything fitting together unobtrusively all the way from the tuck-away headlights to the swept up line at the tail. When in the company of other sports cars, though, the overall size of the Corvette becomes noticeable and, when alongside almost any European sports car the sheer bulk of the Corvette is most apparent.

In many ways, the Corvette is the original "build to suit" sports car. There is a complete range of options that make it possible to satisfy almost any driver who might consider buying such a car. It can be had "mild" with automatic transmission, power steering, power windows or you can have it "wild" with everything up to and including the fuel-injected 375-bhp engine, heavy duty suspension, cerametallic competition brakes, fast steering, wide-base cast-alloy wheels and the whole biz.

CORVETTE COMBINATIONS

bhp	3-speed all-synchro	4-speed all-synchro	Power-glide
250	x	x	x
300		x	x
350		x	
365		x	
375		x	

Note: 250-bhp engine with 3-speed gearbox is standard equipment; all other combinations optional at extra cost.

Because General Motors and the United Auto Workers were agreeing to disagree at the time we had been promised a full-options model, we were not able to obtain precisely the version we most preferred for test. Instead we had to borrow a car from Harry Mann Chevrolet in Los Angeles through arrangements ⟫⟫→

⟫⟫→

CORVETTE STING RAY
AT A GLANCE...

Price as tested............................$5581
Engine.............V-8, ohv, 5356 cc, 375 hp
Curb weight, lb.............................3050
Top speed, mph..............................138
Acceleration, 0–60 mph, sec.................6.3
Passing test, 50–70 mph, sec................2.4
Average fuel consumption, mpg...............13

1965 CORVETTE STING RAY

made for us by Frank Milne for our pictures, and borrow a broken-in model from a private owner for test figures.

Anyone who has driven a Sting Ray will feel right at home in the 1965 Corvette. The new seats are firm but comfortable, the driving position is improved by the telescopically adjustable steering column (3-in. total travel) and the complete instrumentation is a pleasure to contemplate, though reflection from the glass makes them hard to read about half the time. The engine whumps into life with a mean rumble at a twist of the key, the clutch is deceptively easy in operation yet positive in action, the stubby gearshift lever moves into place with a crispness that is pure sensual pleasure. On the negative side, we aren't altogether pleased with the vision to the rear of the coupe. The combination of inside and outside mirrors work pretty well, but looking back over your shoulder (in parking, or taking a quick look for overtaking traffic before a lane change) there is a blind spot that could hide a Cadillac.

Other than this, however, the driving of the Corvette is hard to fault. It is a car that can be driven as the occasion requires. The clutch can be eased out to slip away from a stop with only an almost imperceptible change in the exhaust note. Or you can feed in a lot of throttle and drop the hammer. It's happy either way, which says a lot for the car. Making a fast start, thanks to good weight distribution, independent rear suspension and positraction, the big machine simply squats and squirts. Oh yes, it is possible to apply black stripes of rubber in whatever lengths you like, but it doesn't require a drag race expert to get good, clean, fast starts time after time.

But it was the brakes of the '65 Corvette that pleased us most of all. Long ago we gave up (read chickened out) on doing stomp-down, all-out panic stops in American cars, but the Corvette restored our faith to such an extent that we did 0-80-0-80-0 time after time and grew bored, almost, with the ease and lack of fuss with which the car stopped straight and true. No lock up, no fade, no muscle-straining increases in pedal pressure. Just good dependable stops. Wonderful. And such a great improvement over the standard drum brakes of previous years. We're not knocking the competition cerame-tallics, of course, which are still available (for $628 extra) for the all-out racing Corvette.

It is an open secret that a bigger engine is being prepared for the Corvette and that it is scheduled to be available shortly after the first of the year. This engine, it is said, will be of 396-cu-in. in displacement and will incorporate many of the features of the tremendously impressive 427-cu-in. Chevy stock car engine that appeared (then disappeared) just at the time GM forsook racing two years ago. Though 400–425 bhp is certainly going to propel the competition version along at even higher rates of speed than the present 375, we're not at all certain that more horsepower is the answer so far as competition is concerned. The Corvette is simply too far over-weight when it comes to competition like factory Cobras.

It is the opinion of our design experts that Chevrolet stylists could have made the Corvette look more like a genuine GT car in the "Italian tradition," but that the market for a car of that type is limited (because there aren't all that many people who really care). There's no reason to think that General Motors doesn't know this. GM is pretty successful at designing cars for the people who will buy them.

Summing up, the car does a masterful job of hitting the market bulls-eye. The entire car, and especially the interior, is keyed to the boulevardier sports/racey types who account for the great majority of its sales. It has enough pizzazz for a movie set, or crumpet collecting, or nymphet nabbing, or for the types who get their jollies from looking at all that glitter. It encourages the Walter Mittys to become Fangios or Foyts. Yet it also goes well enough to suit the driver who is sincere about going fast, and can handle that much performance with skill.

CORVETTE STING RAY

SCALE: 10" DIVISIONS

PRICE

List price..................$4463
Price as tested...........$5581

ENGINE

No. cylinders & type.....V-8, ohv
Bore x stroke, in....4.00 x 3.25
Displacement, cc..........5356
   Equivalent cu in........326.7
Compression ratio.........11.0:1
Bhp @ rpm.........375 @ 6200
   Equivalent mph..........132
Torque @ rpm, lb-ft...360 @ 4000
   Equivalent mph...........85
Fuel injection, type.........port
Type fuel required.......premium

DRIVE TRAIN

Clutch type......single plate, dry
   Diameter, in.............10.0
Gear ratios, 4th (1.00).....3.70:1
   3rd (1.28)............4.74:1
   2nd (1.64)............6.07:1
   1st (2.20)............8.14:1
Synchromesh............on all 4
Differential type......limited slip
   Standard ratio.........3.70:1
   Optional ratios........6 ratios
        from 3.08 to 4.56

CHASSIS, SUSPENSION

Frame type: ladder type with 5
  cross members.
Brake type..................disc
  Swept area, sq in........461
Tire size................7.75-15
Steering type....recirculating ball
  Overall ratio...........17.6:1
  Turns, lock to lock.......2.92
  Turning circle, ft..........39
Front suspension: independent with
  unequal A-arms, coil springs,
  tube shocks, stabilizer bar.
Rear suspension: independent with
  fixed differential, lateral leaf
  springs & struts, U-jointed drive
  shafts, tube shocks.

ACCOMMODATION

Normal capacity, persons.....2
Seat width, front, in.....2 x 19.5
Head room.................38
Seat back adjustment, deg......0
Entrance height, in..........49.5
Step-over height..........15.5
Door width, front/rear......31.5
Driver comfort rating:
  For driver 69-in. tall........90
  For driver 72-in. tall........75
  For driver 75-in. tall........70
  (85–100, good; 70–85, fair;
    under 70, poor)

GENERAL

Curb weight, lb...........3050
Test weight................3430
Weight distribution (with driver),
  front/rear, %..........48/52
Wheelbase, in.............98.0
Track, front/rear.....56.8/57.6
Overall length, in..........175.1
  Width..................69.6
  Height.................49.8
Frontal area, sq ft........19.3
Ground clearance, in........5.0
Overhang, front/rear......32/45
Departure angle (no load), deg..18
Usable trunk space, cu ft....10.5
Fuel tank capacity, gal......20.0

INSTRUMENTATION

Instruments: 7000-rpm tach, 160-
mph speedometer, ammeter, water temperature, oil pressure,
fuel.
Warning lights: parking brake,
headlights, high beam, turn indicator.

MISCELLANEOUS

Body styles available: coupe, convertible & hardtop convertible.

ACCESSORIES

Included in list price: 375-hp engine, 4-speed transmission, disc
brakes, special suspension, full
instrumentation.
Available at extra cost: power windows, radio, power antenna,
power steering, air conditioning,
leather seats, telescopic steering
column.

CALCULATED DATA

Lb/hp (test weight)...........9.3
Cu ft/ton mi................155
Mph/1000 rpm (high gear)...21.3
Engine revs/mi.............2812
Piston travel, ft/mi..........1523
Rpm @ 2500 ft/min......4615
  Equivalent mph...........98
R&T wear index...........42.8

MAINTENANCE

Crankcase capacity, qt.........4
  Change interval, mi.......6000
Oil filter type...........paper
  Change interval, mi.......6000
Chassis lube interval, mi...12,000
Tire pressure, front/rear,
  psi.....................24/24

ROAD TEST RESULTS

ACCELERATION

0–30 mph, sec...............2.9
0–40 mph...................3.9
0–50 mph...................5.2
0–60 mph...................6.3
0–70 mph...................7.8
0–80 mph..................10.0
0–100 mph.................14.7
Passing test, 50–70 mph.....2.4
Standing ¼ mi.............14.4
  Speed at end, mph.........99

TOP SPEEDS

High gear (6500), mph.......138
3rd (6500).................110
2nd (6500)..................85
1st (6500)..................64

GRADE CLIMBING

(Tapley data)

4th gear, max gradient, %.....22
  3rd......................31
  2nd......................40
  1st..................off scale
Total drag at 60 mph, lb......135

SPEEDOMETER ERROR

30 mph indicated......actual 30
40 mph.....................40
60 mph.....................60
80 mph.....................79
100 mph....................98

FUEL CONSUMPTION

Normal driving, mpg.......11–15
Cruising range, mi......220–300

ACCELERATION & COASTING

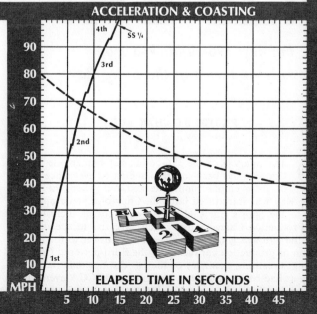

ELAPSED TIME IN SECONDS

425 BHP CORVETTE

*The big new 396-cu-in. V-8 from
Chevrolet is thrust in the Sting Ray*

PHOTOS BY CHAN BUSH

THE ARRIVAL OF the 396-cu-in., 425-bhp engine in the Corvette has been eagerly anticipated ever since word of its eventual coming leaked out when the 1965 Sting Ray was introduced last fall. Chevrolet made application to the Sports Car Club of America for its inclusion in the Production Category list, which gave rise to speculation that the 396 was going to be Chevrolet's answer to the domination of production car racing by the Ford-powered Cobras. Living, breathing examples of the car did not materialize as early as had been anticipated, however, and SCCA eventually removed it from the approved list for 1965. It is available now, though, and will no doubt be eligible for SCCA amateur racing in 1966.

The use of the 396 engine has required only a few changes in the basic 1965 Sting Ray. These changes include an increase in clutch plate pressure, the use of a slightly wider radiator and larger fan, a slightly bigger anti-roll bar at the front and the addition of an anti-roll bar at the rear. The axle shafts and U-joints have also been strengthened by using 4240 alloy steel and then peening them.

The Corvette will now be available with the standard 327-cu-in. engine (there is the choice of 250, 300, 350 or 375 bhp) and with the 425-bhp version of the 396 engine. The models with the "small" 327 engine will still have the full range of gearbox options (3-speed manual is standard, 4-speed manual and automatics are optional) but with the 425-bhp engine only the close-ratio 4-speed is available.

Known as the "Turbo-Jet 396," the new V-8 is a thoroughly up-to-date production version of the "Porcupine" head engine which made a brief but spectacular appearance at Daytona three years ago. It replaces the "Turbo-Fire 409" throughout the Chevrolet line and though there are some points of similarity between the two engines, none of the parts are actually interchangeable.

So far as the Corvette is concerned, the 396 is a consider-

ably beefier engine than the 327. It is a cast-iron engine in the traditional sense (no nonsense about "thin wall" casting techniques here), bigger in physical bulk than the 327 and about 80 lb heavier. The all-up weight of the 396 engine is about 680 lb.

The cylinder heads of the 396 are its most interesting features. These use a modified wedge-type combustion chamber design with the spark plug located in the center for more uniform flame front and the valves are seated at odd angles to provide improved gas flow. The odd angles at which the valve stems are disposed (they are tilted away from each other in both the longitudinal and transverse planes) resulted in the "Porçupine" appellation affixed to the Daytona engine of three years ago.

In all other respects, the 396 engine is traditional and straightforward. The lower end is very hefty, boasting something like 12% more bearing area than the old 409-cu-in. engine and four, rather than two, bolts in each bearing cap. The main bearings of the 396 are 2.75 by 0.992 in. compared with the 2.30 x 0.752 mains of the 327, which gives a general idea of the greater beef in the bigger engine.

Our test car was a convertible and it was a pleasure to make its acquaintance again. The cloth top, which is easily and conveniently stowed under the panel behind the seats, is the very model of what a convertible (or roadster) top ought to be—including admirable weather tightness when that is needed. There's still something just slightly wrong

about the seating position in the Corvette even though the steering wheel is now adjustable for three inches of in-and-out travel. If the steering wheel is pushed in to where it feels comfortable for the arms, there's too little room for the knees to operate the clutch. And if you pull it out to make room for the knees, you have the steering wheel rim too close to the chest. This is assuming that the driver is six feet or over, you understand, as the smaller driver will probably not find fault with the arrangement. There's also still a lot of glare from the glass covering the otherwise admirable instruments (which ⟫→

425 BHP CORVETTE

we've mentioned before). While we're discussing the accessories and accouterments of the Sting Ray, we have been reminded that the heater of the Corvette is almost alone among sports car heaters in that it really works—even in cold weather.

But the 396 engine is the news, right? So what does the 396 engine do for the Corvette? First of all, it brutes it up a bit. When you lay your foot on the high-up clutch pedal, you begin to get the message. This pedal is stiff, hard to depress, cursed with an over-center spring that requires the greatest amount of effort at the top of the stroke where you have the least amount of leverage. Then, after you get the pedal down and select a gear, you find that all the action is concentrated in the final half-inch of the movement on the way out. These two characteristics make it difficult to make a smooth, unobtrusive, while-the-cops-are-watching start. It is perhaps superfluous to add that the clutch is positive in action though, you bet, and there's never any doubt when it is engaged.

The 4-speed all-synchro gearbox, the same as before, is about as near faultless as any we've ever encountered. In obtaining our performance figures at Carlsbad Raceway, we tried a number of different starting techniques. Raising the revs and dropping the clutch, the usual procedure, got us a

lot of wheelspin, a stench of singed rubber and a best standing quarter-mile time of 14.4 sec. Easing off the line with a chirp and then being careful to keep the tires just this side of broken loose gave us a 14.1. This is quick. Quicker than any other standard production car we've ever tested except the AC Cobra. Quick enough that nobody is likely to give you much trouble getting away from a stoplight. Except the law, maybe.

A note must be added here, though, about traction and tires. Our test car had 8.4 lb/bhp and a standing quarter of 14.1. The last Cobra we tested, a 271-bhp street version, had 9.4 lb/bhp but shot through the quarter in 14 flat. This says something about tires and rubber on the road. The tires on the 396 Corvette were 7.75-15 U.S. Royal low-profile "Laredos," the same as those on the standard, less powerful Corvettes and, frankly, the 396 could use something a bit wider and stickier. Given that, there's no doubt at all that you could see the underside of a 14-sec standing quarter.

It is difficult to describe precisely the 425-bhp Corvette's place in the automotive scheme of things. It's an interesting technical exercise, building a nice big engine like the 396 and putting it in a good chassis like the Corvette, but it honestly isn't a very satisfactory car for driving in everyday traffic. It's too much of a brute for that. And with all that power, any manner in which it is driven on anything except dead dry paving, the car is going to be a very large handful. It is not a car for the inexpert or the inattentive—two blinks of the eye and a careless poke of the toe and you could be in serious trouble.

Is the 396 Corvette going to seriously challenge the AC Cobra on the U.S. road racing circuits next year? We wish we could say it would, as we think it would be good for the sport. Unfortunately, however, the Corvette is simply too heavy to make the question anything more than academic. At 3200 lb, the Corvette is roughly 800 lb heavier than the Cobra and this is an impossible amount of weight to be giving away. Increasing the brute horsepower will increase the top speed but the limiting factor in road racing has never been how many horses there are under the hood but how many can be efficiently delivered to the road. If you put two cars through the same turn at the same speed you can predict, other things being more or less equal, that it's going to be the heavier car that slides off the outside of the turn first. And that's the Corvette.

If the 396 version isn't going to restore the glorious name of Corvette to the top of the SCCA racing heap, and we don't think it is, the role of the car becomes somewhat clouded. It seems to us that the addition of the bigger engine is departing from the proper approach. The proper approach, we think, is to make the most efficient use of a good design—which is what the fuel-injected 327 Corvette did—not simply to stuff in a bigger, stronger engine. There are many sports cars that honestly need more power. But the Corvette isn't one of them.

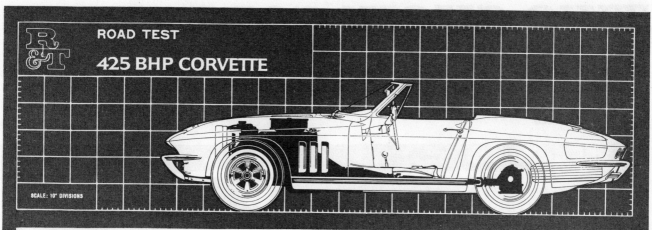

425 BHP CORVETTE

SCALE: 10" DIVISIONS

PRICE

List price...............$4705
Price as tested...........$5185

ENGINE

No. cylinders & type....V-8, ohv
Bore x stroke, in......4.09 x 3.76
Displacement, cc...........6489
 Equivalent cu in...........396
Compression ratio........11.0:1
Bhp @ rpm.........425 @ 6400
 Equivalent mph...........136
Torque @ rpm, lb-ft....415 @ 4000
 Equivalent mph.........85
Carburetors, no. & make..1 Holley
 No. barrels & dia......4V-1.686
Type fuel required......premium

DRIVE TRAIN

Clutch type......single plate, dry
 Diameter, in...........10.4
Gear ratios: 4th (1.00)....3.70:1
 3rd (1.28)............4.74:1
 2nd (1.64)............6.07:1
 1st (2.20)............8.14:1
Synchromesh............on all 4
Differential type......limited slip
 Ratio.................3.70:1
Optional ratios: six, from 3.36 to
 4.56:1

CHASSIS & SUSPENSION

Frame type: box section ladder with
 cross members.
Brake type..................disc
 Swept area, sq in..........461
Tire size.................7.75-15
 Make........U.S. Royal Laredo
Steering type....recirculating ball
 Turns, lock to lock.........3.4
 Turning circle, ft.........39.9
Front suspension: independent with
 unequal A-arms, coil springs,
 tube shocks, anti-roll bar.
Rear suspension: independent with
 lateral leaf spring, lateral struts,
 U-jointed half axles, tube shocks,
 anti-roll bar, trailing arms.

ACCOMMODATION

Normal capacity, persons........2
Seat width, front, in....2 x 20.5
Head room.................39.0
Seat back adjustment, deg.....2
Entrance height, in..........46.5
Step-over height............14.4
Door width.................37.5
Driver comfort rating:
 For driver 69-in. tall........90
 For driver 72-in. tall........85
 For driver 75-in. tall........75
 (85–100, good; 70–85, fair;
 under 70, poor)

GENERAL

Curb weight, lb............3260
Test weight................3570
Weight distribution (with
 driver), front/rear, %....51/49
Wheelbase, in..............98.0
Track, front/rear.......56.8/57.6
Overall length, in..........175.1
 Width....................69.6
 Height...................49.8
Frontal area, sq ft.........19.3
Ground clearance, in........5.0
Overhang, front/rear......32/45
Departure angle (no load), deg..18
Usable trunk space, cu ft......8.0
Fuel tank capacity, gal......18.5

INSTRUMENTATION

Instruments: 7000-rpm tachometer,
 160-mph speedometer, fuel, ammeter, oil pressure, water temperature.
Warning lights: parking brake,
 headlights retracted, high beams,
 turn signals.

MISCELLANEOUS

Body styles available: convertible
(as tested), convertible with hardtop and coupe.

EXTRA COST OPTIONS

Power windows, power brakes,
power steering, knock-off wheels,
air conditioning, hardtop, radio,
fiberglass tank, heavy duty suspension.

CALCULATED DATA

Lb/hp (test wt)..............8.4
Mph/1000 rpm (4th gear)....21.2
Engine revs/mi............2820
Piston travel, ft/mi.......1770
Rpm @ 2500 ft/min........4000
 Equivalent mph.........85
Cu ft/ton mi...............181
R&T wear index...........49.9

MAINTENANCE

Crankcase capacity, qt.........6
 Change interval, mi.......6000
Oil filter type..........full flow
 Change interval, mi.......6000
Chassis lube interval, mi.....6000

ROAD TEST RESULTS

ACCELERATION

0–30 mph, sec...............3.1
0–40 mph...................3.8
0–50 mph...................4.8
0–60 mph...................5.7
0–70 mph...................7.5
0–80 mph...................8.9
0–100 mph.................13.4
50–70 mph (2nd gear).......2.5
Standing ¼-mi, sec.........14.1
 Speed at end, mph........103

TOP SPEEDS

High gear (6400), mph.......136
 3rd (6400)................106
 2nd (6400)................83
 1st (6400)................62

GRADE CLIMBING
(Tapley data)

High gear, max gradient, %....22
 3rd.....................28
 2nd.....................38
Total drag at 60 mph, lb.....125

SPEEDOMETER ERROR

30 mph indicated....actual 27.4
40 mph....................37.6
60 mph....................57.8
80 mph....................77.4
100 mph...................98.0

FUEL CONSUMPTION

Normal driving, mpg.......9–12
Cruising range, mi.......165–220

ACCELERATION & COASTING

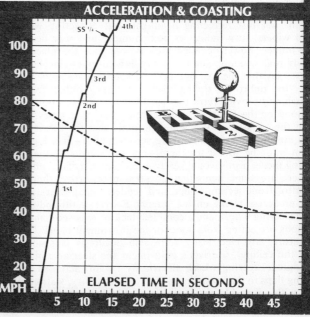

ELAPSED TIME IN SECONDS

TWO SHOW CORVETTES

*Two very different directions the Corvette might
go—and what'll you bet they don't take the wrong one?*

BY RON WAKEFIELD

The Mako Shark II: an exercise in surface development that is more commercial than significant.

Ever since the present Corvette Sting Ray series went into production almost three years ago, we have cherished a little glimmer of hope—that the next Corvette would really "arrive" as an exemplary sports/GT car. The production Sting Ray comes close: it has an admirable chassis layout, replete with fine suspension, weight distribution, steering, brakes (now) and instrumentation. But it is too heavy, a bit too large, and overstyled to a degree that dulls its appeal to the discerning enthusiast. Now Chevrolet gives us a new Show Car which they hint might be an indication of Corvettes to come, as well as another car seen at a minor show without any fanfare or even comment.

The first, called the Mako Shark II and shown for the first time at the 1965 New York Automobile Show, features such delectable items as "fire frost" finish, simulating the coloring of its namesake shark and changing from black at the roofline to dark blue on the sides to light gray at the bottom; not just retracting headlights but retracting windshield wipers, turn indicators, rear bumper and rear license plate holder; power-operated louvers that take the place of a lowly rear window; and the heavy 396-cu-in. engine. Press releases say that the rear bumper and license plate holder retract "for unbroken styling lines." But they didn't follow through and provide retractors for the various projections on the hood or the array of grids, chrome bars and pipes that adorn the car's sides. We would presume that the retracting items are intended for improving aerodynamics—if a logical reason must be found for them—but the weight of the retracting mechanisms would surely cancel any beneficial effect.

There are some items that do seem noteworthy, though: an entire front section—hood and fenders—that raises as a unit for access to the engine, Jaguar E-style; a tilt-and-telescope steering wheel, and a complete lack of exposed handles or knobs projecting from doors or the instrument panel. Other features are electrically adjustable headrests (suspended from the ceiling), a driver's adjustment for steering ratio, a rectangular steering wheel, electrically (again) actuated stabilizing flaps in the rear, flip-up roof for easy entry, stationary seats with adjustable pedals, and a plethora of warning lights and gauges. Length of the Mako is 9.5 in. longer than the

production Sting Ray, height 3.5 in. lower, and the width the same. Weight of the car or details of its body construction have not been revealed.

The design of the body and the "styling" of it are, as so often with GM cars, two separate concepts. The basic lines are pleasing and exciting—squint your eyes and see. But styling gimmicks and details have been heaped upon it in such abundance that it's really difficult to see the lines. We suppose that this treatment is some kind of entertainment for the masses. Entertainment it is, in the same vein as comic books or pornography. No esthetic experiences here!

The second of the show cars is a competition Sting Ray roadster seen at a recent show at Notre Dame University (South Bend, Ind.). It appears to be a cut-down version of the lightweight competition coupes that appeared all too briefly in mid-1963, only to be canceled by a "re-affirmation" of GM's compliance with the AMA anti-racing pact.

Roadster at Notre Dame—not pretty, but functional.

Changes from the competition coupe, other than the removal of the top, include flared wheel openings to cover larger tires, a raised hood section to provide clearance for the stacks on four dual-throat Weber carburetors, opened-up front fender vents and a built-in headrest for the driver. The cockpit seems to be unchanged. Some of these alterations to the car, such as the wheel-well flares, came about while two of the cars were in active competition—the two cars released by GM after the project had been canceled. The Webers are new to us, as the first examples had fuel injection.

This car represents the type of development we're interested in seeing for the Corvette. Weight has been saved by various means. The standard box-section frame was replaced by one made of aluminum tubing; the body was formed by making a hand layup of thin glass cloth instead of the spray-nozzle application of chopped fibers used in production; and the engine has an aluminum block. The interior uses the standard instruments and steering wheel, and probably some weight has been saved in the instrument panel area. The effect is not nearly as stark as that of many such cars, however, and the cockpit looks quite well finished as well as purposeful.

The overall weight of the coupes and probably this roadster, 1870 lb, represents of course an extreme in weight-saving for racing purposes. But much of what is used here could be moderated in cost and still be feasible for reducing the weight of the production car. For instance, the tubular frame could be made of steel instead of aluminum and probably still save a lot of weight as compared with the production one.

Even though this car's concept antedates that of the Mako Shark by two years, we think that the roadster is a much more significant car. Our faith in Arkus-Duntov's engineering team tells us that future Corvette series will represent real progress, not just gadgetry.

Lightweight competition car as it appeared in 1963.

1967 CORVETTE

Unique among American cars—and among sports cars

THE STING RAY is in its fifth and probably last year with that name and body style, and it finally looks the way we thought it should have in the first place. All the funny business—the fake vents, extraneous emblems and simulated-something-or-other wheel covers—is gone, and though some consider the basic shape overstyled, it looks more like a finished product now.

Actually, the car has undergone few changes in its 5-year run. The suspension and sound deadening were refined in the 2nd year, disc brakes introduced in the 3rd and the needlessly large "porcupine" engine added as an option in 1966. Minor refinements have been made all along; sales have risen gradually each year. And it remains unique among American cars —and among sports cars.

For this test we selected an almost basic version of the Corvette: a convertible hardtop with the standard 300-bhp 327 engine, 4-speed gearbox and power assists. With the exception of the power assists, we concluded that this is the Corvette for the thinking driver.

The test car was certainly impressive looking. Its paint was better than on any Corvette we've tested, a deep claret red. Its restrained trim and the new wide-rim (6 in.) wheels gave it a purposeful, almost elegant look.

Performance can be taken for granted with over 5 liters and 300 bhp in a 3160-lb car. It is a credit to the American way of doing things that the performance is achieved in a remarkably silent, smooth and economical manner. Our test car had the optional 4-speed gearbox, which is smooth, quiet and easy shifting (the linkage has been isolated from the vibration that was an annoyance in earlier models); but now that the standard 3-speed has been redesigned and includes a synchronized 1st, we can see no good reason for ordering the 4-speed for general use. Other than the ease of reselling, that is.

The only thing that marred the flexibility of performance in our test car was the unwillingness of its engine to return to idle speed. For one thing, its throttle linkage was a bit sticky; but more basic than that is the fact that the anti-backfire valve used with emission-controlling air injection dumps a blob of fresh air into the intake manifold for about 2 sec after the foot is lifted, causing the engine to hesitate momentarily at 1200 rpm on its way down to the normal 500-rpm idle speed. Minor, but bothersome, for California customers.

It's hard to find fault with the Corvette's handling; it's as near neutral as any car we know and of course there's always enough torque available to steer with the throttle. It is quite stable directionally in a cross wind and imparts an immense feeling of security out on the open road. Steering effort with the standard 20.2:1 overall ratio is moderate, and the customer may have the ratio lowered to 17.6:1 by a simple repositioning of the steering linkage and resetting of the toe-in, if he's willing to accept parking efforts in the Armstrong category.

Our test car, however, was equipped with the optional linkage power steering; and in our opinion this system is unacceptable. Like other GM power steering it is set to begin assisting the driver only after he has exerted a pre-set amount of force at the wheel rim. On the Corvette, this takeover-point apparently has been raised—most likely to give more "feel." But the result is almost the opposite. The wheel takes an almost constant amount of effort (about 7 lb) for any maneuver. Corvette is the last GM car to retain linkage-assist power steering and it's reasonable to predict that next year it will have the integral unit used on other GM models—and hopefully with more accuracy.

Over smooth roads the Corvette has a pleasing, almost big-car ride. Noise level is low from most sources, and thus this is a car for long-distance travel. Its effective heater, excellent AM/FM radio and good weather protection (regardless of top ordered) place it near the top of the list of sports/GT cars for pure comfort. On rough roads the ride deteriorates as the progressive-rate springs, limited wheel travel and willowy body structure combine to put up quite a fuss. Under these conditions it is no match for the best European examples, like the Mercedes 230 SL or BMW 2000 CS. But then this is More Car for the Money, and perhaps it is unrealistic to expect ultra-sophisticated chassis design. In all fairness, however, this is America's best chassis.

The Corvette's interior abounds with niceties. There are a handy new center-mounted handbrake, seat-belt hooks that solve the problem of where to stow them, large, readable instruments and handy controls. Materials are nice but not extravagant, and if the whole effect is one of overstyling, at least the layout is logical and comfortable. An adjustable steering column is optional, and seatbacks can be adjusted over a small range with a screwdriver and wrench. New seatback locks keep them upright until they're released for access to the adequate, but awkward, luggage area behind.

Disc brakes on all four wheels are powerful, smooth and

consistent. Constant contact of the pads with the discs permits a very small travel of the caliper cylinders and thus allows a large mechanical advantage in the actuating circuit: because of this the Corvette is the only production car over 3000 lb with four discs that can get by without power brake assist. And though our test car's power assist gave adequate feel, we consider this item to be unnecessary. Fade characteristics of these Delco-Moraine brakes are excellent, but their ultimate stopping capability is limited to 24 ft/sec/sec or 0.75-g by tire adhesion. The front brakes locked consistently and evenly at this deceleration rate. Squeal prevails when the discs are cool. Parking brake actuation is by small drum units built into the rear discs; in our experience this arrangement works well, but poor adjustment on the test car limited its holding power to about 10% grades.

One unwelcome change from earlier Sting Rays is the amount of engine fan noise in the current model. Apparently nothing has been changed but the shape of the fan shroud, but the fan even with its viscous drive is a veritable windmill all the time. Perhaps the change has achieved the greater air flow ⟫→

1967 CORVETTE
AT A GLANCE

Price as tested	$4824
Engine	V-8, ohv, 5356 cc, 300 bhp
Curb weight, lb	3160
Top speed, mph	121
Acceleration, 0–¼ mi, sec	16.0
Average fuel consumption, mpg	15.4

Summary: Performance & handling with comfort, reliability & longevity. Rather large & heavy for 2-only seating because of separate body/frame. Good value for money. Probably last year for this body style.

New for 1967 are four-lamp flasher control on steering column, center handbrake, console placement of fresh-air vent controls.

1967 CORVETTE

needed with the air injection smog pump to provide adequate cooling under idling conditions.

All things considered, the Sting Ray is a big value for the money. It matches any of its European competition for useful performance and walks away from most of them; it's quiet, luxurious and comfortable under ordinary conditions; easy to tune and maintain; and even easy on fuel if its performance isn't indulged too often. Quality of assembly is lacking, however, and the following items were amiss on our test car: several rattles; improper clutch adjustment; an air leak over the windshield; choke setting; sticky throttle linkage; and a fresh-air vent that wouldn't shut off.

The improvements we would most wish for in the next Corvette series would be lighter weight, improved body structure and quality control, and a better ride on poor surfaces. But in the meantime the Corvette ranks with the best sports/GT cars the world has to offer, regardless of price. It is significant to note that with the power assists omitted, the Corvette tested can be bought (through the normal haggling process) for approximately $4000. 🌀

Emission-reducing air pump is on right front of engine. Wire isn't standard—accordion hose was an emergency replacement.

The handsome new 6-in.-wide standard wheel. Power steering plumbing can be seen inboard, just under frame rail.

ROAD TEST
1967 CORVETTE

SCALE: 10" DIVISIONS

PRICE

Basic list.................$4228
As tested.................$4824

ENGINE

Type.........................ohv V-8
Bore x stroke, mm.....102 x 82.6
  Equivalent in.......4.00 x 3.25
Displacement, cc/cu in..5356/327
Compression ratio........10.25:1
Bhp @ rpm..........300 @ 5000
  Equivalent mph.............111
Torque @ rpm, lb-ft..360 @ 3400
  Equivalent mph..............78
Carburetion........one Holley 4-V
Type fuel required.......premium

DRIVE TRAIN

Clutch diameter, in..........11.0
Gear ratios: 4th (1.00)......3.36:1
  3rd (1.46).............4.90:1
  2nd(1.88).............6.32:1
  1st (2.52)............8.47:1
Synchromesh............on all 4
Final drive ratio.........3.36:1
  Optional ratios..........3.08:1

CHASSIS & BODY

Body/frame....steel ladder frame,
  separate fiberglass body.
Brake type: vented discs, 11.75-in.
  diameter; single calipers.
  Swept area, sq in........461
Wheel type & size....disc, 15x6JK
Tires.....UniRoyal Laredo 7.75-15
Steering type....recirculating ball
  Overall ratio..........17.6:1
  Turns, lock-to-lock........2.9
  Turning circle, ft........41.6
Front suspension: independent with
  unequal-length A-arms, coil
  springs, tube shocks, anti-roll bar.
Rear suspension: independent with
  lateral leaf spring, lateral struts,
  U-jointed halfshafts, trailing
  arms, tube shocks.

OPTIONAL EQUIPMENT

Included in "as tested" price:
  AM/FM radio, 4-speed gearbox,
  limited-slip diff, power steering
  & brakes, exhaust emission control, minor items.
Other: various engine options, automatic transmission, air conditioning, power windows, competition
  equipment.

ACCOMMODATION

Seating capacity, persons.......2
Seat width ....2 x 19.0
Head room.................39.0
Seat back adjustment, deg......2
Driver comfort rating (scale of 100):
  Driver 69 in. tall...........90
  Driver 72 in. tall...........85
  Driver 75 in. tall...........75

INSTRUMENTATION

Instruments: 7000-rpm tachometer,
  160-mph speedometer, fuel level,
  ammeter, oil pressure, water
  temperature.
Warning lights: parking brake, brake
  fluid loss, headlights retracted,
  high beams, directional signals.

MAINTENANCE

Crankcase capacity, qt.......5.0
  Change interval, mi......6000
Filter change interval, mi....6000
Chassis lube interval, mi.....6000
Tire pressures, psi........24/24

MISCELLANEOUS

Body styles available: convertible
  with hard and/or soft top, coupe.
Warranty period, mo/mi.24/24,000

GENERAL

Curb weight, lb 3160
Test weight 3540
Weight distribution (with
  driver), front/rear, %....49/51
Wheelbase, in............98.0
Track, front/rear.......57.6/58.3
Overall length...........175.1
  Width.................69.6
  Height................49.8
Frontal area, sq ft.........19.2
Ground clearance, in........5.0
Overhang, front/rear....32.0/45.1
Usable trunk space, cu ft.....8.0
Fuel tank capacity, gal......18.5

CALCULATED DATA

Lb/hp (test wt).............11.8
Mph/1000 rpm (4th gear).....23.1
Engine revs/mi (60 mph)....2600
Piston travel, ft/mi.........1405
Rpm @ 2500 ft/min.......4610
  Equivalent mph..........103
Cu ft/ton mi..............139
R&T wear index............36.6
Brake swept area, sq in/ton...261

ROAD TEST RESULTS

ACCELERATION

Time to distance, sec:
0–100 ft 3.5
0–250 ft 6.1
0–500 ft 9.0
0–750 ft 11.3
0–1000 ft 13.4
0–1320 ft (¼ mi)........16.0
Speed at end of ¼ mi, mph...86.5
Time to speed, sec:
0–30 mph 3.4
0–40 mph 4.6
0–50 mph 5.9
0–60 mph 7.8
0–70 mph 10.0
0–80 mph 12.9
0–100 mph 23.1
Passing exposure time, sec:
  To pass car going 50 mph...4.9

FUEL CONSUMPTION

Normal driving, mpg...14–18
Cruising range, mi....260–325

SPEEDS IN GEARS

4th gear (5500 rpm), mph.....121
3rd (5500).................85
2nd(5500) 68
1st (5500)................50

BRAKES

Panic stop from 80 mph:
  Deceleration, % g........75
  Control............excellent
Fade test: percent of increase in
  pedal effort required to maintain
  50%-g deceleration rate in six
  stops from 60 mph.........16
Parking: hold 30% grade.......no
Overall brake rating.....very good

SPEEDOMETER ERROR

30 mph indicated......actual 29.0
40 mph.................39.4
60 mph.................59.3
80 mph.................78.1
100 mph...............95.6
Odometer, 10.0 mi.....actual 9.8

ACCELERATION & COASTING

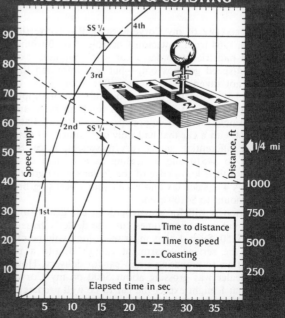

Speed, mph

SS ¼   4th

3rd

2nd   SS ¼

1st

Distance, ft

¼ mi

1000

750

500

250

Time to distance
Time to speed
Coasting

Elapsed time in sec

5   10   15   20   25   30   35

AFTER THE NEW WEARS OFF:
CORVETTE STING RAY, 36,000 MILES LATER

BY RON WAKEFIELD

I TOOK DELIVERY of my Corvette Sting Ray convertible on June 19, 1964 from Patterson Chevrolet Co., Birmingham, Mich. I had tried to order a car with minimum extras—radio and 4-speed gearbox only—but was informed that orders were not' taken for Corvettes after about June 1 of a given model year because of the close relation of supply and demand. So I shopped a few dealers for both price and the ability to supply me a car equipped as I wanted it. Patterson came closest with a car that had only one additional item, the limited-slip differential, and a very reasonable price. Sticker price came to $4510 and I, dealing as a normal customer (albeit with a little knowledge of the dealer's cost) got the car for $3825. A 4% sales tax and a half-year's licensing brought the total amount to $4100 even.

Competition is stiff for a metropolitan car dealer, and the pressure to cut prices sometimes cuts the dealer's *gross* profit to around $150. Such dealers must sell a high volume of cars, and often the service department cannot devote a proper amount of time to preparing a new car for delivery. However, since the dealer was grossing about $400 on my Corvette, I reasoned that I might expect the car to be well prepared. The car was delivered to me with no more than a wash job, apparently. Problems: a broken spark plug; the side windows wouldn't roll up with the doors shut; splotchy paint on top cover panel; the top didn't fit properly; the passenger's seat rattled furiously; leaks around the windshield; and the steering had a massive squeak in it.

Details of the experience of getting these faults corrected approach the sordid and don't bear going into here; since then I've come to the conclusion that the condition of my car was typical and the dealer service was typical; in my experience it seems most Chevrolet dealers look upon the Corvette as some kind of Funny Furrin Car.

The next shock was insurance. I was at the time working off a few minor traffic violations (Detroit police don't like MGs using their ability to maneuver in traffic), and premiums run rather high on Corvettes to begin with. I had passed age 25, fortunately, and got by for $374 the first year. Later the rate came down.

The rest of the story for the first 36,000 miles is more pleasant. I finally readjusted the convertible top myself; the weatherstripping in the top for sealing the door windows couldn't be adjusted sufficiently to allow the windows to roll up easily, but I must say that the windows do seal well now! The other problems were taken care of by the dealer in his good time.

At 7500 miles the car was pulling slightly to the left, and

I had the front end aligned and wheels balanced for $12.95 before embarking on a cross-country trip. During the trip, at 8209 miles on the odometer, the tach suddenly fell to zero and the engine lost half its power. The tachometer drive gear in the distributor base had jammed and stripped itself, and in the process jerked the distributor around, throwing the timing off. A Louisville, Ky., dealer didn't have the necessary gear but did reset the timing and point gap for $3.50. I completed the trip without a tachometer and later had this repaired on warranty.

The long, hard Michigan winter brought out both the best and worst in the Corvette. I had been driven to buy this rather large, heavy car because the climate forced me to appreciate its powerful heater (a string of British sports cars made me appreciate heaters) and its rustproof plastic body—the salt mines are on overtime up there in the winter. Cold starting had something to do with it too, for none of my British jobs could be coaxed to start at much below zero F. Sure enough, the Corvette started instantly all winter; all that was necessary was one poke of the throttle before turning the switch. And that heater—wow! After a few miles at 15 below the blower could be cut back to its lowest (of three) speed and the temperature control could be reduced from maximum. And it was certainly nice not to worry about those cakes of salt on the bodywork.

That was the best part. The worst was the performance of the Positraction limited-slip differential on glazed-ice roads. This option is just fine for getting one out of the mud or off the line at the drag strip, but when both wheels are in sub-marginal traction conditions, the limited-slip can't make up its mind which should have the torque, and the car proceeds down the road mildly fishtailing from side to side. Under these circumstances, all the little old ladies in their Valconbler II sedans drove right past me and my white knuckles. The original-equipment tires supplied by Chevrolet didn't help things either, but tightwad that I am, I was determined to wear them out before replacing them with something more suitable.

The owner's manual for the Corvette suggests an oil and filter change and chassis lubrication every 6000 miles, and an engine tune-up every 12,000 miles, with the usual admonitions about doing these things more often under adverse conditions. Because most of my driving is such that I warm up the car thoroughly on each trip, I stuck to the maximum intervals on the lubrication and oil changing. I had the first engine tune-up—points, plugs, condenser and carburetor adjustment—done at 13,791 miles by a Chevrolet dealer for $33.85. Not that the engine needed it badly; but I was still proud of the new car and wanted to give it the best.

The winter brought on another problem. Where the clutch linkage enters the bell housing, there is a rubber boot. The boot broke (and has regularly broken since), allowing the salty slush into the clutch where it rusted the throwout bearing solid. The symptom of this condition was that a binding could be felt in the clutch pedal when the engine was running, but not when it was shut off. The dealer had some difficulty in figuring this out—as did I—and replaced the bearing on warranty after finally diagnosing the trouble. In the meantime, I had had the boot replaced at a cost of $3.65. The bearing was replaced at 19,555 miles.

At 20,000 miles a loud "pop," heard on applying or releasing power, developed in the rear end somewhere. Two dealers weren't able to find the cause, and one did replace the bracket which secures the differential to the frame. This didn't solve the problem, and to date I still have it. I now think that it originates as a binding in the limited-slip unit, and hope to report on this later.

It took me less than a year to use up the 24,000-mi warranty, and 24,000 miles were passed with no regrets except for the popping differential. As the engine was quite smooth and peppy, I decided to postpone the specified tune-up for a while.

The original U.S. Royal 800 tires may not have been star performers, but they wore well and evenly and required replacement at 29,500 miles. A new set (four) of 205-15 Pirelli Cinturatos set me back $277, but the newfound security in the rain made them seem well worth it. I selected the 205-15 size because of its approximation to the rolling radius of the original 6.70-15s, but in retrospect I think 185-15s would be better for the car. The left front tire rubs a frame rail when cranked hard over to the left, and steering effort is greater than it would be with the smaller size. Ride harshness, already great in the Corvette, is worsened by these tires and getting them properly balanced has been a problem. Still, all things considered, they are good for the car. I retained the original cross-ply spare, and from experience can say that it's absolutely *verboten* to run any distance on three radials and one cross-ply.

At 30,093 miles it was necessary to straighten the left strut rod in the rear suspension for some unknown reason; the dealer (in California) straightened it because there was no replacement rod in stock. Realignment of both front and rear suspension was also performed at this time, plus wheel balancing all around. Total cost for these operations was $34.20.

At 34,000 miles the throttle pedal became arthritic and I replaced it for $1.90. At 35,000 the cool-side tailpipe rusted through and was replaced for $15.86. It should be noted here that I installed a set of straight-through mufflers, given to me as a gift, a bit earlier; otherwise there might have been a muffler replacement in the repair totals.

Oil consumption was a little heavy at first (800 mpq) and it took some heavy-footed driving to get the rings seated. Since then, the usage rate has been about one quart per 1400 miles. Fuel economy for the relatively mild 250-bhp engine has been agreeable: 18.1 mpg overall, corrected—the odometer reads 102.3 miles for every 100 traveled. The engine does, however, require premium fuel.

Certain observations are in order to explain the rather low operating cost for the period. For instance, at 36,000 miles the engine still didn't need a tune-up; also, because I do most of my own washing, polishing and de-rattling, there's very little money spent for this kind of maintenance. And as the car was not quite two years old when it turned 36,000 miles its depreciation for the mileage is relatively low ($3000 value as of May 1966). CONTINUED ON PAGE '61 ⟫⟫

CORVETTE STING RAY

Repairs & Replacements in 36,000 Miles

Change steering to fast ratio, reset toe-in at 915 mi....$	9.00
Replace broken spark plug, correct seat rattle, adjust weatherstripping and top at 1816 mi.............warranty	
Align front end, balance two wheels at 7500 mi........	12.95
Reset timing and points at 8209 mi (failed tach drive)..	3.50
Replace tachometer drive.........................warranty	
Tune engine at 13,791 mi........................	33.85
Replace rubber boot, clutch linkage at 18,469 mi.......	3.65
Replace clutch throwout bearing at 19,555 mi........warranty	
Replace differential bracket at 23,629 mi............warranty	
Tires, 4 new at 29,500 mi........................	277.00
Straighten rear suspension rod, align both ends, balance all wheels at 30,093 mi....................	34.20
Repaint nose, damaged by stone....................	6.00
Replace throttle pedal at 34,000 mi..................	1.90
Replace tailpipe at 35,000 mi......................	15.86
Total repairs and replacements in 36,000 mi.........$397.91	

PHOTO BY BILL MOTTA

CORVETTE STING RAY, 1963-67
Fast, reliable sports cars for the budget-minded enthusiast

OVER ITS TWENTY-three years of being America's only locally produced sports car, the Corvette's place among its international competitors has waxed and waned. It got off to a rather halfhearted start in 1953 as a two-seater with Chevrolet sedan suspension components, a 6-cylinder engine and automatic transmission—only an automatic transmission. Things got better in succeeding years: the hot new Chevrolet V-8 engine became available in 1955, an altogether more handsome body came with the 1956, and fuel injection and a 4-speed gearbox made the 1957 a landmark car. Although the 1958 was improved again in functional matters, its gaudy new styling details weren't so nice, and matters became worse when the 1961 got a new rear end that was distinctly unharmonious with the front.

The 1963 Corvette, however, was except for its engine and transmission options an almost all-new car. Gone was the now outdated (in such a high-powered car) live rear axle, replaced by new independent rear suspension; the front suspension became 1963 Chevrolet instead of 1953; and the new body and frame were both smaller and lighter. The 1963 represented real progress over any previous Corvette. That basic Corvette continues today, though with a much larger but less roomy body that first appeared with the 1968 model.

To the serious shopper for used sports cars, this middle category of Corvettes—1963 through 1967—offers the greatest

potential. The 1963–67 Sting Rays were good, straightforward sports cars, getting better each year as Chevrolet tidied up both the styling and the engineering. Two R&T editors owned Corvettes of this series, one a 1963 and the other a 1964, and both look back on their cars with fondness. Granted, time dims the problems and enhances the pleasant experiences, but both objectively recall that the good times far outnumbered the bad. Of course those "middle period" Corvettes weren't perfect, but they were handsome and comfortable, had plenteous V-8 power and offered a long list of options that allowed one to tailor the car to one's own tastes. And they were fun to drive. If you shop carefully today it's possible to have all this in a used Corvette at a fraction of the price of a 1975: indeed, at less than a 1975 4-cylinder sports car.

If you are like most longtime enthusiasts you have probably been on one side or the other of the First Corvette argument. For the record, there *were* 1953 Corvettes, 300 of them to be exact. If in your search for a used Corvette you stumble on a 1953, buy it, put it in your living room and plan on sending your kids through college with the money from its sale later.

The Corvette was General Motors' move to capitalize on the still small but burgeoning sports car market in the U.S. In 1952 just slightly more than 11,000 new sports cars were registered in this country, the majority of them MG TDs. In Detroit, where new-car sales had gone past the 4-million-yearly mark, 11,000 anythings weren't very significant. But Harley Earl, head of GM Styling, liked sports cars and was "secretly" at work with a few members of his styling staff on a two-seater he figured GM could sell in sufficient quantities to make a profit on it. With a series of well timed moves Earl got the approval and support of first one, then another top GM executive and the first Corvette was previewed at the 1953 Motorama, a road show GM then took around America each year. The rest, as they say, is history. Although Ford countered with the two-seat Thunderbird in 1955, by 1958 the T-Bird had become a four-passenger personal luxury car and the Corvette again had sole possession of the "only American sports car" title.

Of the new 1963 Corvette, a special supplement to Chevrolet's *Corvette News* said, "In the last few years Corvette demand has exceeded supply; so from the standpoint of popularity an entirely new vehicle was not necessary. Nevertheless, we felt that the original design no longer represented our best engineering, so plans for a change were initiated in 1959." The same situation applies today and it's still unknown just when the next all-new Corvette will appear, but that's another story.

The new Corvette was offered in both roadster and coupe versions, and it appeared larger than its predecessor despite its smaller dimensions. But more important than the new styling were the changes beneath the new shape. The independent rear suspension, for instance, was straightforward and effective: there were simple lower links, tubular shock absorbers and, surprisingly, a single transverse leaf spring mounted at its center to the differential housing. The axle halfshafts themselves did the work of top lateral links, and trailing arms took braking and acceleration loadings. Front suspension was also much improved. The Elliott kingpin system, actually dating back to the 1949 Chevrolet, was replaced by a spherical balljoint design basically the same that had been used on Chevrolet's passenger cars since 1955. There was a new, more precise steering system that permitted a change in overall ratio from 19.6:1 to 17.0:1 by simple relocation of the tie-rod ball studs on each steering arm. For the first time, power steering was a Corvette option. What had been an optional wheel width (5½ in.) became standard. There were also changes in seating position, pedal location, heating and ventilation—you name it. Our sister publication at the time, *Car Life*, gave the car its Award for Engineering Excellence that year, and our own summation was that "in its nice, shiny new concept it ought to be nearly unbeatable."

The 1963 Sting Ray, then, was something of an engineering *tour de force*, especially for the cost-conscious domestic car industry. And if the 1963 wasn't quite perfect—it still had drum brakes, there were unpleasant body resonances and the shift linkage buzzed—it was to be refined steadily over the next four years under the able guidance of Zora Arkus-Duntov, acquiring better engineering and shedding styling excesses.

The 1964 version benefited from very small refinements. The fake grilles in the hood were the first excrescences to go, although the indentations for them remained. The divided rear window in the coupe was replaced by a single, full-width rear window. (Ironically, the split rear window, which was the least-liked styling feature of the first Sting Ray and made rearward vision a problem, now makes the 1963 model very desirable among Corvette enthusiasts.) There was better sound dampening, the chronically noisy shift lever was tamed somewhat and the ride was made less harsh with variable-rate springs.

For 1965 there was a greater leap forward: four-wheel disc brakes were made standard. Also, a 396-cu-in. V-8 joined the 327 that had been standard Corvette fare since 1962. The 396, offered only that one year, developed 425 horsepower (gross) at 6400 rpm, more power than had ever been available in a Corvette, and added about 200 lb of weight. This year the hood indentations were smoothed out (although 396-engine cars had a "power dome") and fake body-side vents became real.

The next two years were basically refinement years, but some dramatic engine changes are worthy of note. It was the brute-performance era. Great throbbing gobs of horsepower and torque, that's what it was all about in those halcyon pre-⟶

1964 coupe: revised roof vents and wheel covers, hood grids removed.

1965: real front-fender vents, new wheel covers again, smooth hood.

Year of stability: most noticeable 1966 change was new wheel covers.

Final cleanup: new wheels, new vents, fewer badges identify the 1967.

The vision-hindering 1963 split rear window has acquired rarity value.

Independent rear suspension was introduced on the 1963 Sting Ray.

The 427 engine was much bulkier and heavier than the standard 327.

PHOTO BY CHAN BUSH

News. We have no figures on the production of L88-equipped Corvettes, but then you don't really want 560 bhp. Do you?

The market for used Corvettes is in a curious state. In some parts of the country they are much sought after, rare and expensive. In other sections little attention is paid to them. For example, in California used Corvette prices are about 25 percent *lower* than in the eastern states, just the reverse of the situation with imported sports cars. Bob Wingate, a Southern California trader in Corvettes, volunteered some interesting information.

According to Wingate, there are very few Corvettes still in good, unaltered (not customized) condition. "Fifty percent of all Corvettes ever made," Wingate explains, "have been stolen at one time or another. Of that 50 percent about 75 percent are gone for good, after being stripped for parts or what have you. Take away the Corvettes that have been customized or turned into race cars of various types and it is estimated only 20 percent of pre-1968 Corvettes are still available in basically good, stock condition."

It's those clean, stock Corvettes that are appreciating by 20 percent a year and it's those cars that are so sought after by collectors. So, strange as it may seem, a used Corvette can be a good investment. Where else can you earn 20 percent on $3000 in a year?

energy-crisis pre-emissions-crunch days. The 396 grew to 427 cu in.—a full 7 liters—and plain old cubic inches replaced fuel injection as the route to power. In 1965 the injection 327 engine had been a $538 option. The new 427, which generated 425 bhp at 6400 rpm vs 375 at 6200, cost only $313 additional.

The 1967 Corvette was originally scheduled to get the current body style, but as final decision time approached it was obvious the new design still carried some flaws that couldn't be worked out in time so the original Sting Ray design was retained for one more year with minor cosmetic changes. This was the year the body finally looked as it should have all along—most of the identification badges were gone, the wheels were now handsome ventilated discs without fake knockoff hubs and there wasn't a false scoop or vent left. Two 427 engines with three 2-barrel carburetors were added to the option list, one with hydraulic lifters and one with mechanical. The mechanical-lifter version was conservatively rated (unusual in those days) at 435 bhp at 5800 rpm. There was also a special-order 427 with aluminum heads: the legendary L88 option, first offered in the spring of 1967. According to Karl Ludvigsen in his book *Corvette, America's Star Spangled Sports Car, The Complete History,* "There was no hedging on the output of this engine; Chevy didn't say anything about it at all." However, Ludvigsen goes on, reliable reports said 560 hp at 6400 rpm on 103-octane fuel. The L88 engine could be ordered only with all the competition options—plus one option called C48: deletion of the otherwise standard heater and defroster, "to cut down on weight and discourage the car's use on the street," according to *Corvette*

Now we come to what to look for, other than what one checks in any used car. These second-generation Corvettes were the answer to the street racer's dream, so look for obvious signs such as dump tubes (exhaust cutouts) or welded patches on the head pipes that might indicate dump tubes were once installed. Check the U-joints for looseness. There are six of them, one on each end of the driveshaft and one on each end of the two rear axle halfshafts. They take a beating from all that Corvette torque, especially if the car has gone through life at the hands of a competitive driver or drivers. Likewise, the Positraction (limited-slip) differential that is on most Corvettes can suffer from frequent hard driving. To check it, drive around a very tight turn, applying and releasing power gently. If the Positraction chatters, graunches or pops it needs attention. Make sure the noise-suppressing metal shields around the distributor, coil and plug wires have survived: without them the radio is worthless. And of course check for clutch slippage. The easiest method is to hold the brakes on firmly and rev the engine while slowly releasing the clutch with the transmission in 2nd gear. If the engine doesn't quickly stall, the clutch is slipping. Watch the exhaust for blue smoke too; Chevrolet V-8s aren't especially easy on oil.

Cosmetically and esthetically there are some items to note. Flared fenders, three taillights, mag wheels and the like detract from the car's value, so don't pay top money for a car that's modified. On the other hand, don't be scared off by hairline cracks in the paint, but if they go into the fiberglass look underneath for patching that would indicate an incident. And don't be bothered much by rattles and squeaks. They are standard equipment, especially in the convertible, and a little patient work can eliminate a lot of them—as it did monthly on the Editor's 1964.

Without much fanfare, early Sting Rays have caught the collectors' fancy, so what you pay will depend on a great many factors: your geographical area, the condition of the car and such. Any price guidelines we could give would have to be considered very general indeed. For example, a 1963 coupe in restored but not concours condition could range between $3500 and $4200; add $500 for fuel injection and another $500 for the handsome factory-optional aluminum wheels. Luckily, since they are the better cars, 1964-67 Corvettes in good, clean stock condition are less expensive: anywhere between $2000 and $3500, with condition determining price more than model year. And if you don't mind the work you can probably find a tatty example for much, much less and have some good fun bringing it back to life.

With all the engine options that were available over the five years this series was built, it's no surprise that the character of a 1963-67 Corvette can range from quiet, docile GT car to rip-snortin' sports car or near-dragster. Even the mildest 327—rated at 250 bhp by the old SAE gross system but probably something like 180 bhp by today's more conservative method—has a lot of performance by today's standards, though, and with the 4-speed gearbox can do 0-60 in about 8 seconds or cover 18 miles per gallon depending on how it is driven. The Powerglide automatic transmission, though it's only a 2-speed, did pretty well in the Corvette with the ample power and light weight, and it's worth noting that from 1963 on the Corvette's factory air conditioning was better than that of some of the latest, most expensive European sports cars. The 4-speed manual gearbox, on the other hand, is a delight to use, especially in the later models where its characteristic "buzz" was reduced or (in 1967) completely eliminated. You won't find many 3-speeds, but if you do, expect to get a bargain price. Though the 3-speed is perfectly adequate, it's not in demand.

The Corvette Sting Ray series, then, is another prime example of what we're looking for in this Used Car Classic series: an older car that in many ways matches new cars costing a lot more—perhaps even outdoes them, such as in performance and fuel economy. It is entirely possible to have, in a 1963-67 Corvette, a car that lives up to contemporary standards in ride and noise level, comes close to matching them in handling and probably exceeds them in performance and economy—all for less than $4000. ▼

Soft top of the Sting Ray convertible is supremely easy to fold or raise.

The headlights are raised by electric motors with their own switch.

PHOTO BY BILL MOTTA

1963-67 CORVETTE BRIEF SPECIFICATIONS

Curb weight, roadster	3050
Wheelbase, in.	98.0
Track, front/rear	56.8/57.6
Length	175.1
Width	69.6
Height, convertible	48.1
Fuel capacity, gal.	18.5
Engine type	ohv V-8
Bore x stroke, in., 327	4.00 x 3.25
396	4.09 x 3.76
427	4.25 x 3.76

Transmission types: 3-speed manual, 4-sp manual (wide or close ratios), 2-sp automatic

Front suspension: unequal-length A-arms, coil springs, tube shocks, anti-roll bar

Rear suspension: lower lateral links, halfshafts as upper links, trailing arms, transverse leaf spring, tube shocks; anti-roll bar with 396 or 427

Brakes: 1963-64, all-drum; 1965-67, all-disc, vented rotors

Tires & wheels: various sizes bias-ply tires on steel or aluminum 15 x 5½ or 15 x 6 wheels

PERFORMANCE DATA
From Contemporary Tests

	1963 327 FI	1964 327/300 auto	1965 396/425	1966 427/425	1967 327/300
0-60 mph, sec	5.9	8.0	5.7	5.7	7.8
0-100 mph	16.5	20.2	13.4	13.4	23.1
Standing ¼ mi	14.9	15.2	14.1	14.0	16.0
Avg fuel economy, mpg	12.5	14.0	10.5	12.0	16.0
Road-test date	10-62	3-64	8-65	8-66*	2-67

*Car Life magazine. A set of Corvette road-test reprints is available from R&T's Reader Service Dept for $3.25 plus 50¢ postage charge per order. The set includes 15 tests, from the 1954 6-cyl to a 1970 454 automatic. The Car Life test is not included. All tests here are for 4-speeds except the 1964 Powerglide car.

CORVETTE ENGINE OPTIONS
cu in./gross bhp

1963	1964	1965	1966	1967
327/250	327/250	327/250	327/300	327/300
327/300	327/300	327/300	327/350	327/350
327/340	327/350	327/350	427/390	427/390
327/360*	327/365	327/365	427/425	427/400
	327/375*	327/375*		427/435
		396/425		

*fuel injected

Corvette Sting Ray

1965 FUEL-INJECTION CORVETTE *High point*
for the American sports car

PHOTOS BY JOE RUSZ

THE 1965 FUEL-INJECTION Corvette was the technical high-water mark for the whole Sting Ray series. Until that model year Corvettes still had drum brakes and one had to pay more than $600 extra to get a fancy set of drums and real stopping power. On the other hand, 1965 was the last year for fuel injection; from 1966 on Chevrolet got Corvette power by adding cubic inches. So in the 1965 there was a fortuitous coming-together of technologies and for one year America produced a sports car with fuel injection, all-independent suspension and all-disc brakes.

It seems funny, doesn't it, in the context of 1975 to remember that a leading manufacturer had airflow-controlled continuous-flow fuel injection and dropped it? The Rochester system was just that, and though it was quite different in detail and designed for a lot of power rather than low emissions, it answers the same basic description as the system now used on Volvos and Porsches, for instance, to meet 1975 emission regulations. How times change.

Paul and Mary Jacobson of Mission Viejo, California loaned us their bright blue 1965 FI Corvette for this test. It is a prize, as the photos show: they bought it new and had driven it only 38,363 miles when Mary brought it to us. It looks better than new, thanks to some nice chroming Paul has had done under the hood, and performs like the thoroughbred we all know the injection Corvette to be. Naturally the Jacobsons didn't want the car subjected to our usual battery of performance, braking and handling tests, so most data given here is taken from tests done by R&T and our companion magazine at the time *Car Life*. But we were allowed to drive the Corvette as much and as fast as we wanted to, and it was indeed a treat to feel the eagerness of an engine that was designed first and foremost to go.

Look at those specifications. A bit over five liters (the good old 327) . . . 11.0:1 compression ratio! . . . 375 bhp @ 6200 ➤➤

rpm. Okay, gross horsepower they were. But these days 300 bhp net, which is about what this engine developed, would be staggering. The most powerful 1975 Corvette gets by with 205 net—and it weighs at least 300 lb more than the 1965.

Driving the injected Corvette reminded us that fuel injection meant something altogether different in 1965 from what it does in 1975. From our recent experience with injected engines we might have assumed the FI Corvette would be remarkably docile, idle at a low speed and pull strongly from 1000 rpm in 4th. Not so. This is a pushrod, not an overhead-cam, engine and that means lots of valve overlap to get high output. No doubt this engine was more tractable than its 365-bhp carbureted counterpart, but the injected Corvette has the character of a semi-racing engine by contemporary standards. The Jacobson car idles at 1300 rpm and not all that smoothly; looking back at the *Car Life* test we find that the 1964 car idled "with a lope" at 1100. And if we cracked open the throttle sharply at, say, 1500 rpm in any gear the engine was likely to bog down a bit. It felt like a classic case of over-carburetion—something we haven't experienced in a long time.

But once the revs begin to build up, hang on! By the time the big tach needle reaches 3000 rpm, things are beginning to happen so fast it's dizzying. The fuel-injection cars always had the close-ratio 4-speed gearbox; this car had the numerically high 4.11:1 final drive ratio, which makes the most of the close ratios. Even with the 4.11 it takes a good bit of clutch slipping to get the car off the line and it's a nice long climb to the redline at 54 mph. But after that the driver works hard just to keep up with the engine, so fast does the redline come up in 2nd and 3rd, and it takes very little time to redline the engine in 4th gear either for that matter. The pull begins to fall off above 5500 rpm, but it is a brilliant show up to that point.

Extracting all this performance—0-60 in 6.3 sec, 0-100 in a flat 15—is great fun. The Corvette gearbox and shift linkage are in a class by themselves: a strong, unbeatable set of gears and synchronizers and a crisp, light, short-throw lever to control them. There was a tendency in the early cars for the linkage to develop a "buzz" as the miles piled up, but we found none of this in the Jacobson car.

Fuel economy? Well, that's another matter. Tests of the injection Corvette indicate about 13.5 mpg in daily use. The Jacobsons have done 17 mpg on a trip, which is probably about tops. No, it was not designed for economy. We really weren't even thinking much about fuel economy in those days. Quietness isn't one of the car's virtues either. These Corvettes had two big reverse-flow mufflers in their dual exhaust system, which suppressed any outright display of exhaust note in the old sports-car tradition but didn't keep a tinny sort of power sound from coming through. This, the mechanical tappets and the 4.11 gearing conspire to make the Corvette plenty noisy—by 30 mph it has already reached the 70-dBA mark inside. At highway speed the engine sounds as if it's really working and the usual wind leaks around poorly matching windows and hardtop weatherseals contribute their part to the general cacophony. If you wanted a GT car you bought a 250-bhp Corvette with the 3.36:1 or 3.08:1 axle.

The Jacobson car has neither power steering nor power brakes. Very few Corvettes are sold these days without both of these assists, but then they're heavier and the tires are much, much wider. From 1963 on it has been possible to alter the Corvette's non-power steering ratio from 3.4 turns lock-to-lock to 2.9 just by changing from one attachment point of the steering arms to the tierod balljoints and naturally this had been done on the test car. Even at that the steering isn't unduly heavy and it feels precise as well as being plenty quick. But the bias tires used in the early 1960s—or even the new bias-belted Uniroyals Paul had put on the car just before turning it over to us—follow every groove or ridge in the road and remind us how much better radials are. It took the Corvette people a long time to get around to radials.

We weren't able to test the Corvette for cornering power but would estimate it to be capable of about 0.7g on the 100-ft skidpad, about the level of a small sedan on radial tires in 1975. Tires are the big difference, of course, because the suspension can hardly be faulted in the matter of cornering. It can be faulted on ride, though; at low speeds the car jiggles constantly, and yet going over those gentle freeway undulations it has the feeling of being underdamped as the front end floats up and down. Despite the all-independent suspension and how advanced it was in 1963 or 1965, the Corvette's combination of ride and handling isn't impressive by today's standards even if you allow for the skinny tires.

The driver sits low in the Corvette cockpit, bolt upright in seats with fixed backrests; the angle of the entire seat assemblies can be varied with the seat mounting bolts but the range is small. The Jacobson car has the optional telescopic steering column, which helps get the wheel a bit farther from the driver than the standard, very close position for the large wheel. Another option on the car we drove was the teak wheel rim, a $48 item for which the owner had to wait an additional 15 weeks after ordering the Corvette. We'd never seen one before, so these must indeed be rare. But what an improvement over the standard fake-wood rim.

By 1965 flat dial faces had replaced the original Sting Ray plastic cones, thankfully, and the result was an esthetically pleasing and highly legible set of instruments—huge speedo and tach, four small gauges, a trip odometer (extremely rare in American cars) and a large clock in the center of the double-cowl-effect dash. Heater controls (and air conditioning if ordered) consist of pull-and-twist knobs just under the clock and are odd but simple. And effective: no imported car of the time and few today can match this Corvette's powerful heater and capable air conditioning. Another wonderful Americanism in this cockpit are the lightning-fast electric window lifts, which cost just $59 when the car was built. We'd suggest that European designers take a look at them, but if they haven't seen them by now they probably just aren't interested. Perhaps the tighter weatherseal demanded by the Europeans is one reason why their lifts are so slow, however.

The convertible top of this Corvette was another area where it set the standard for other open sports cars to follow. It never quite fit as installed at the factory but there were adjustments here and there and a patient owner could eventually get it snug everywhere. Snug or not, it was marvelously easy to put up and down, folding neatly into the compartment behind the seats after the cover panel was released by a single lever and going into place just as handily with two latches at the windshield header and two pins dropping into holes in the cover panel. The hardtop is another matter, requiring a lot of bolting and unbolting to install or remove, and one assumes the body designers saw it as an all-winter proposition.

What a pleasure it was to try this 10-year-old injection Corvette! Jacobson has not only kept it looking good and in top mechanical condition, but he regularly tends to the rattles that so quickly develop in Corvettes and hence it was tight and solid feeling. All that muscle was never practical, given the restrictions on American driving that were almost as intimidating then as they are now, but at least there were Nevada and Montana without speed limits when this machine was built. With a numerically lower final drive ratio this would have been a Nevada roadburner; in the form we drove its character is more dragster-like and there was a 4.56:1 final drive to make it even more so.

Come to think of it, aside from the sound basic design, this variety of options was one of the Corvette's main attractions in those days. In any form it was a true sports car, able to hold its head up among the world's best, but its personality could be varied by large or small increments with the selection of well conceived engine, transmission and suspension options. With all due respect to the wider tires and wheels of today's Corvette, we still prefer the overall effect of this earlier Sting Ray series and hope that with the next new model the Corvette people will be able to give renewed attention to making Corvettes small and light as well as fast.

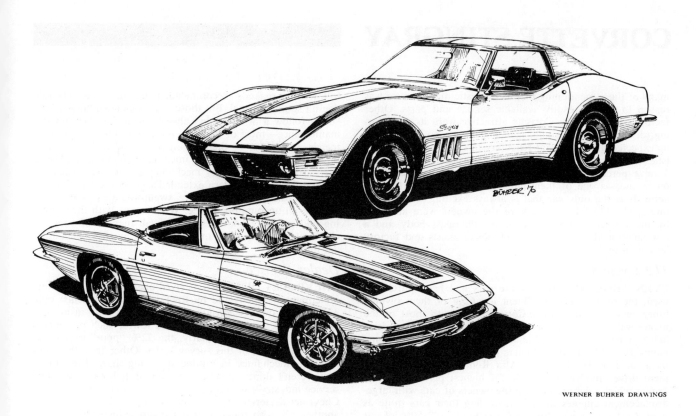

WERNER BUHRER DRAWINGS

Road & Track Owner Survey

CHEVROLET CORVETTE STINGRAY

THE CHEVROLET CORVETTE has been America's only sports car for so many years that its only challengers—the Nash-Healey and Ford's 2-seat Thunderbird—are, respectively, forgotten and a classic. All along it has been something of a stepchild at Chevrolet, being assembled at a truck plant down in St. Louis, but it has had its moments of glory—such as when the all-new Sting Ray was introduced in late 1962—and its sales have continued to grow at a rate that never would have been predicted even by those in Chevrolet Division who fought for getting it into production back in 1963. It has always offered a lot for the money by sports-car standards and at its best it's a fine car to drive.

For months we sent questionnaires out to Corvette owners all over the U.S. and Canada—and even a few to Europe. Of those returned by owners of 1963-69 Corvettes, 177 were sufficiently filled out with the needed information. There were 114 owners of the original 1963-67 Sting Ray series and 63 of the revised 1968-69 Stingrays; average odometer mileage for the early series was 37,500 and for the new series 13,000. Highest mileage reported was 120,000 and 16% of the cars had more than 50,000 miles on their odometers—a tribute to the longevity of the major mechanical components of Corvettes. Distribution of the cars in the survey by engine was as follows: 327-350 cu in./250-300 bhp, 78 cars; 327-350 cu in./340-350-365 bhp, 53 cars; 427 cu in./390-400 bhp, 30 cars; and 427 cu in./425-435 bhp, 16 cars. None of the relatively rare 1965 396s was included.

The Corvette Mystique

THE PRESTIGE of owning a Corvette is indisputable, and many of our respondents mentioned that "girls like it" or "Corvettes are the greatest." But only 7% of them actually said that the "image" was one of their reasons for buying a Corvette. Styling—perhaps another way of saying "image" —was, however, the leading reason for purchase, being specified by 52% of the owners; this isn't nearly as high a percentage as that for the Jaguar E-type, though. The big V-8 engines and their torquey performance were next, 46% of the owners being influenced by them. Thirty-three percent mentioned handling as a reason for getting a Corvette. Then there's the fact that the Corvette is an American-built sports car: 17% of the owners gave this as a reason for purchase,

85

CORVETTE STINGRAY

many of them mentioning that its domestic manufacture was reason to expect readily available service and parts. Other factors that made the Corvette attractive were its engineering (disc brakes all around, independent suspension—14%), its expected durability and reliability (an assumed characteristic of American cars) and the value for money spent. Twelve percent of the owners had owned Corvettes previously and, as one of them put it, "This is my fourth Corvette. It is the only car that is always exciting to drive for under $10,000." Finally, 8% of the owners were attracted by the no-rust characteristic of the fiberglass body and a few mentioned the availability of power assists and factory air conditioning.

The Owners

EVEN THOUGH 24% of these cars had been purchased used, the owners are an affluent lot, 62% of them owning other cars as well as their Corvettes. As with most of the makes we've surveyed, the greatest proportion of them drive between 10,000 and 15,000 miles per year, but their driving habits resemble those of Jaguar or MG drivers more than those of Porsche, BMW or Alfa Romeo drivers: 44% of them drive "moderately," a tie for highest percentage with Jaguar which might tell us that the owners of cars with large-displacement engines don't have to flog their cars quite as hard. Certainly if one drives a Corvette "hard" (50% of our owners do, though only 6% said they drive "very hard") one is courting trouble with the police. Perhaps they do most of their hard driving in organized competition: the healthy percentages of 29% participate in rallies and 18% in slaloms.

Some most interesting things come to light about how the Corvette owners take care of their cars. They're by far the most fanatical group of owners we've come across when it comes to this: the table doesn't show it, as the 62% who follow the maker's recommended maintenance schedule doesn't sound unusual. But those who responded "mostly" or "don't follow it at all" didn't always mean that they did less maintenance—most of them do *more* than the book recommends, fully 33% of them changing the oil and filter more often than the recommended intervals of 6000 miles! We've been around enough older mechanics and drivers to sense a certain distrust of modern extended service intervals and the Corvette owners seem to have swallowed this "old wives' tale" to an uncommon degree. We disagree with them, having had perfectly good results with all manner of cars, including Corvettes, doing maintenance at much greater mileage intervals.

Dealer Service

CHEVROLET DEALERS received one of the most scathing indictments yet delivered by R&T survey respondents: only 32% of the owners consider their dealer "good" in his service dealings, this being the second lowest such percentage we've ever recorded. Similarly, the 32% "poor" rating was the second worst in this survey series. Other than the comments on mechanics' ineptitude and being charged for things that weren't done—practices typical of today's automobile service industry—we found 7% of the owners telling us that Chevrolet dealers aren't familiar enough with Corvettes and another 7% who found that parts were difficult to get. One problem may be that Chevrolet dealers, with their high volume and generally popular-priced cars, may be the wrong outfit to be handling a specialized car like the Corvette: as one owner put it "Although the car costs $5000 they treat you with the same disregard as if you bought a $2000 Chevy II." Another, an ex-GM management employee, said, "I am compelled to state that dealer service countrywide is abominable, with few exceptions."

SUMMARY: CORVETTE STINGRAY OWNER SURVEY

New or Used
Bought new......76%
Bought used......24%

Miles per Year
0-5000...........6%
5000-10,000......14%
10,000-15,000....40%
15,000-25,000....34%
Over 25,000.......6%

How Owners Feel About Corvette Service
Rated "Good".....32%
Rated "Fair"......28%
Rated "Poor".....32%
No opinion........8%

About Driving Habits
Owners who said they drive "Moderately"..44%
Owners who said they drive "Hard"......50%
Owners who said they drive "Very Hard"....6%

Factory Maintenance Schedule Followed?
Owners who followed schedule completely..62%
Owners who followed it mostly...........25%
Owners who didn't follow it at all........13%

Problem Areas Mentioned by more than 10% of the owners:
Instruments
Cooling system
Engine mechanical (427)
Body parts

Rain leaks
Windshield wiper cover (1968-69)
Carburetor

Mentioned by between 5 and 10% of the owners:
Shift linkage
Wheel alignment
Clutch

Wiring
Alternator, regulator
Gearbox

Owners reporting no problems............7%

How Many Current Corvette Owners Would Buy Another?
Would............81%
Would not........15%
Undecided........4%

Five Best Features
Handling
Performance
Styling
Brakes
Reliability, Durability

Five Worst Features
Workmanship
Rattles & Squeaks
Luggage Space
Interior Heat
Ride

"Best" and "Worst" Features

BUT TO THE cars. The things Corvette owners like best about their cars are, as usual, about the same things that induced them to buy the cars in the first place: handling (61%), performance (60%), styling (28%), brakes (26%) and reliability-durability (18%). Other items highly favored by the owners were comfort, with several comments on the Corvette's powerful heater-defroster unit, and the sometimes surprisingly good fuel economy that goes with the smaller engines.

The worst thing about Corvettes, according to the owners, is the workmanship—or the lack of it. Fully 26% of these Corvette drivers found the lack of quality control in assembly the worst feature. This we expected from previous experience; we have found that the cars are usually delivered to the customer with many things wrong, the new owner spending his first few weeks getting the car sorted out rather than enjoying it. Also, confirming what we'd found in 1968-69 test cars, the quality level of the new series is significantly worse than it was for the 1963-67 models, presumably as a result of increased demand for, and production of, the new style. Whereas 18% of the owners of 1963-67 models considered workmanship the worst feature, an astounding 40% had that comment to make about the 1968-69 models!

Rattles and squeaks were No. 2 on the "worst" list. To a degree these are directly related to the quality-control troubles, but they are also inherent in the Corvette's construction; its fiberglass body lacks rigidity, most of the required strength being furnished by the heavy steel frame underneath, and as a result the body is prone to creak and shake. The convertible models are, of course, worse than the closed ones; the 1963-67 coupe was relatively free of body noises. Seventeen percent of the owners of both series listed squeaks and rattles as the worst feature.

Luggage space was next on the worst feature list, 11% of the owners complaining of either too little space or the inaccessibility of it. Then came ride, 7% of the owners noting its harshness. Fourteen percent of the 1968-69 owners complained of poor ventilation for the interior, and although they were not always specific about having air conditioning, we believe two factors need mentioning in this connection: the big engines—especially 427s—generate a lot of underhood heat that finds its way inside the car, and when the present model is ordered with air conditioning its fresh-air ventilation system is sacrificed. Two "worst features" that weren't mentioned in significant proportion, but which are nevertheless part of the Corvette story, made clear by our owners' comments were high insurance rates and the likelihood of the car's being stolen!

Component Life and Problem Areas

THE LARGE number of respondents in this survey enabled us to pin down the life of certain normal replacement items comparatively accurately for some, but not all, of the models included in the survey. These are summarized at right. Tire life was greater for the pre-1968 models which did not have "wide oval" tires; life of the disc brakes used from 1965 on does not appear to be much different from that of the drums used before. Clutch life is as expected: shorter for the "hot" or high-output version of an engine vs its "mild" counterpart because the former won't have as much low-speed torque and will therefore require more clutch slippage to get the car moving from rest. Harder all-around usage is a factor too with the hot engines. As for the spark plugs, the high-output engines need new plugs more often because at low speeds their large valve overlap periods cause poor combustion and results in large deposits on the plugs.

There was only one engine overhaul due to wear reported, and that, at 45,000 miles, for a high-performance 427 used in Germany. However, there was a high incidence of engine mechanical problems occurring early in the life of 427s, generally having to do with the valve gear.

Getting to the problem areas, the greatest one was the instruments. In particular, 15% of the owners had trouble with the speedometer, odometer or cable; 13% with the tachometer or its mechanical drive from the distributor; and 12% with the minor instruments, the clock being the worst offender. Then there was the cooling system, a problem for 20% of the owners. Radiator, hoses and just plain overheating were among the specific problems noted, and one particular design slipup—that of the alternator belt rubbing the top hose on 1967 models—stood out (we'd also had this happen to us on our 1969 427 test car). One owner told us the solution for his 1967 327/300: use a longer belt, part no. 3847707 from a 1962 Corvette, and the top hose from the 1965 Corvette 327, part no. 387366.

Next most frequent problem was body parts, with a 15% trouble rate on bits and pieces from fiberglass cracks to window regulators; 13% of the 1968-69 owners had trouble with the automatic hatch cover for the windshield wipers. Rain leaks were noted by another 15% of all owners, and the carburetor was a sore spot for 12%—or 19% of the 1966-67 models, the first years of Holley carburetors. Among those problems occurring for 5-10% of the owners, wheel alignment caused premature tire wear and we tabulated it as a problem where the owner noted either that his new car was delivered incorrectly aligned or that alignment (front or rear) was a recurring problem. For those reporting routine alignment only, the mileage interval was 23,000 on the average. Speaking of tire wear, 11% of our Corvette drivers had fitted radials and were getting somewhere around 30,000 miles tread life from them.

Summary

THAT'S THE CORVETTE, then—the only car of its kind built in America, an exciting thing to drive and behold but rather a letdown when the eager new owner finds out how poorly it is assembled. So how do the owners feel about their Corvettes when asked the crucial question, *Would you buy another?* Eighty-one percent would, 15% wouldn't and 4% are as yet undecided. Here's how that compares with other sports and GT cars covered in our surveys so far:

	Corvette	Alfa	Datsun	Jaguar	MGB	Porsche	Triumph
Would	81	94	80	83	70	95	67
Would not	15	3	18	17	19	4	30
Undecided	4	3	2		11	1	3

"Poor quality" was the only reason directly related to the owners' cars that occurred in any significant number as a reason not to buy another, but a very interesting thing did crop up from the owners of the older series: 17% of these people wouldn't buy the present model because they find it unattractive for one reason or another—its larger size and increased gadgetry, mainly. But obviously the fact that these people won't buy another isn't bothering Chevrolet as more and more people scramble to buy new Stingrays.

AVERAGE COMPONENT LIFE

Component	Miles	Component	Miles
Tires		Clutch	
(1963–67 models)	18,500	(mild 327–350)	57,500
Tires (1968–69)	11,900	Clutch (hot 327–350)	38,000
Shock Absorbers	28,600	Sparkplugs	
Brakes, drum	37,000	(mild 327–350)	10,800
Brakes, disc	38,800	Sparkplugs	
Mufflers		(hot 327–350)	6,700
(standard exhaust)	29,000	Sparkplugs (mild 427)	5,600

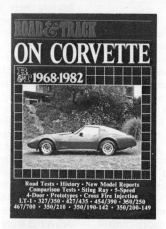

ROAD&TRACK
ON CORVETTE 1968·1982